D0871758

COMPUTATIONAL
WELDING MECHANICS

 Springer

COMPUTATIONAL WELDING MECHANICS

by

John A. Goldak
Professor of Mechanical & Aerospace Engineering
Carleton University, Ottawa Canada

Mehdi Akhlaghi
Associate Professor of Mechanical Engineering
Amirkabir University of Technology, Tehran Iran
Visiting Professor at Carleton University, Ottawa Canada

 Springer

Library of Congress Cataloging-in-Publication Data

A C.I.P. Catalogue record for this book is available
from the Library of Congress.

ISBN-10: 0-387-23287-7   e-ISBN 0-387-23288-5   Printed on acid-free paper.
ISBN-13: 978-0387-23287-4

Printed in the United States of America.

9 8 7 6 5 4 3 2 1   SPIN 11054023

springeronline.com

Table of Contents

Chapter I: Introduction

Chapter II: Computer Simulation of Welding Processes

Chapter III: Thermal Analysis of Welds

Chapter IV: Evolution of Microstructure Depending On Temperature

Chapter V: Evolution of Microstructure Depending On Deformations

Chapter VI: Carburized and Hydrogen Diffusion Analysis

Preface

The presentation of this book emphasizes an exposition of the computational principles and the application of computational welding mechanics to practice.

The objective in Computational Welding Mechanics is to extend the capability to analyze the evolution of temperature, stress and strain in welded structures together with the evolution of microstructure. Distortion caused by volumetric strains due to thermal expansion and phase transformations are a dominate load in the stress analysis. The microstructure evolution influences the constitutive equations. In particular, as the temperature changes from above the melting point to room temperature, the stress-strain relationship changes from linear viscous, to visco-plastic to rate independent plasticity. In high strength steels, transformation plasticity can have a major affect in reducing the longitudinal residual stress in welds.

In the past twenty years the capability to analyze short single pass welds including the above physics has been developed. The software engineering to develop codes to solve these coupled problems has tended to be near the limit of what users and software developer can manage. The complexity has been too great to develop codes to deal with long multipass welds in complex structures. To deal with this problem, new software engineering methods and strategies have been developed, that are able to automatically create initial conditions, boundary conditions, adaptive mesh generation and manage time stepping. The input is a functional specification of the problem that includes a weld procedure, a weld path and a structure to be welded. The weld procedure contains weld parameters for each weld pass in a weld joint.

By giving designers the capability to predict distortion and residual stress in welds and welded structures, they will be able to create safer, more reliable and lower cost structures.

In the next ten years, we predict that computational weld mechanics will be used routinely in the welding industry. We believe that adequate solutions now exist to solve the problems related to managing the complexity now exist of software and the ease of use of the software.

There are some longer term issues that will require further research. One issue will be obtaining values (preferably functions) for the temperature and history dependent material properties of base metal and weld metal. Another issue will be obtaining sensor data characterizing welds and welded structures. We anticipate that real-time computational weld mechanics will lead to computer controlled welding systems that utilize a model not just of the weld pool but of the structure being welded. We also expect that this will lead to archiving huge amounts of data characterizing the welding process of constructing large structures such as a ship or submarine. Some of these archives will be public and some will be corporate. With such archives, data mining will be used to optimize processes and designs. In all of these endeavors, computational weld mechanics will play an essential role. In that role, the fundamental principles upon which computational weld mechanics is based will not change.

It is our hope that this book will help the reader to learn, understand and apply computational weld mechanics to problems in industry.

John A. Goldak and Mehdi Akhlaghi

Units

Currently changes are being made in the use of units from the English system to the *SI* system (le System international d'Unites). However, many articles referred to in this book use the English system and many readers of this book are still accustomed to the English system.

Below is a conversion table for units frequently used in this book.

To convert from	to	multiply by
inch *(in.)*	meter *(m)*	2.54×10^{-2}
foot *(ft)*	meter	3.048×10^{-1}
lbm/foot³	kilogram/meter³	1.601×10
Btu	joule *(J)*	1.055×10^3
Calorie	joule	4.19
lbf (pound force)	Newton *(N)*	4.448
pound mass *(lbm)*	kilogram *(kg)*	4.535×10^{-1}
Lbf/inch² *(psi)*	Newton/meter²	6.894×10^3
kgf/meter²	Newton/meter²	9.806
Fahrenheit (t_F)	Celcius (t_C)	$t_C = (5/9)(t_F - 32)$

Notations

Efforts were made to use the same notation symbols, as much as possible, to express various quantities throughout the entire book. However, the authors have found that it is almost impossible to use a single, unified system of notation throughout the entire book because:

1. The book covers many different subjects.
2. The book refers to works done by many investigators who have used different notations.

Efforts were made to provide sufficient explanations whenever symbols are used.

Chapter I

Introduction

1.1 Introduction and Synopsis

Welding as a fabrication technique presents a number of difficult problems to the design and manufacturing community. Nowhere is this more evident than in the aerospace industry with its emphasis on performance and reliability and yet where materials are seldom selected for their weldability.

Developments in calculating the thermal cycle and elastoplastic stress-strain cycle have been slow because of the inherent complexity of the geometry, boundary conditions and the nonlinearity of material properties in welding.

However, the exponential growth in computer performance combined with equally rapid developments in numerical methods and geometric modeling have enabled computational weld mechanics to reach the stage where it can solve an increasing number of problems that interest the industry specially in pipelines, power plants, refineries and pressure vessels, nuclear reactors, building and bridges, automotive, trucks and trains, ships, offshore structures, aerospace structures, micro electronics and many others.

Although the ability to perform such analyses is important, the real justification for computational weld mechanics is that it is becoming cheaper, faster and more accurate to perform computer simulations than to do laboratory experiments. Taken to the extreme all relevant decisions would become based on computer simulations. For example since nuclear testing in the atmosphere has been banned, this has actually occurred in nuclear weapons design, a field at least as complex as welding. It is unlikely that computational weld mechanics will eliminate all experiments in welding. Instead, computational weld mechanics is likely to increase the demand for accurate constitutive data, particularly at high temperatures, and to include the effect of changes in microstructure. Also it will not eliminate the need for experiments that simulate or prototype processes and products. However, it will dramatically reduce the number and cost of such experiments and greatly enhance the accuracy and significance of the data obtained for each experiment. In the automotive industry, *CAE* (Computer Aided Engineering) is said to have reduced the number of prototypes required from a dozen to one or two.

In the next few years, digital data collection of not only welding experiments but of production welding will be coupled to computer models. The computer models will use the experimental data to adjust the parameters in the computer model. The experiment and production system will use predictions from the computer model to control the process. The mathematics of this is called a Kalmann filter. In this case, the computer model and experiment are tightly connected. Neither could exist without the other. At this point religious wars between experimentalists and theorists will become meaningless.

Furthermore, models can be examined to provide insight that could never be obtained by experiment. For example, it is well known that work piece distortion caused by welding austenitic stainless steel is some three times greater than that caused by welding carbon steel. By analyzing models in which each property is varied separately, the sensitivity of the distortion to each property can be computed. This would provide the knowledge needed to understand the greater distortion in austenitic steel. Of course, this is not possible experimentally.

In its narrowest sense computational weld mechanics is concerned with the analysis of temperatures, displacements, strains and stresses in welded structures, Figure 1-1.

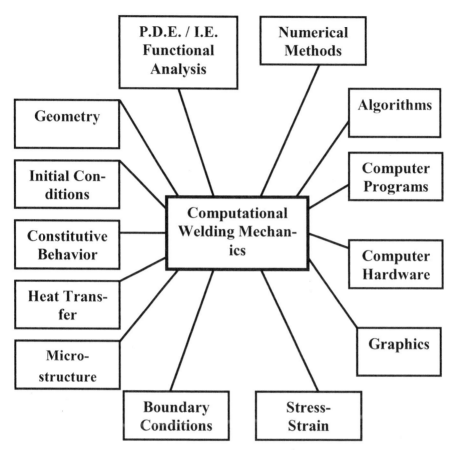

Figure 1.1: Computational welding mechanics draws on the disciplines shown above to compute the temperature, microstructure, stress and strain in welds. (PDE/IE-stands for Partial Differential Equations/ Integral Equations)

In its broadest context, it is an important element of Computer Aided Design *(CAD)* and Computer Aided Manufacturing *(CAM)*. Computer modeling, in general provides the capability of storing vast amounts of data; of organizing and storing relations between data in databases or knowledge bases; and of using these data to compute or predict the behavior of products, processes or systems in the real world. On one hand, it can be viewed as a set of analytical tools for determining the mechanical response of a work piece to a given welding procedure. On the other hand, it can be viewed as a design tool for predicting the quality of a weld and the deformation, Figure1-2.

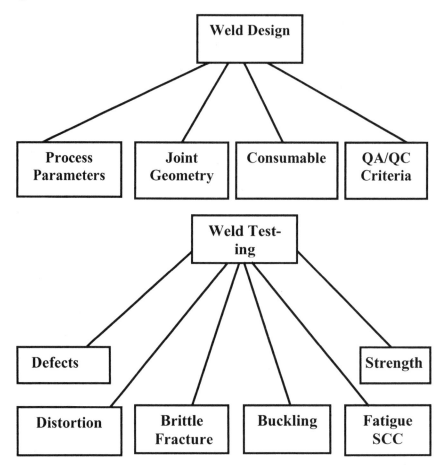

Figure 1-2: The important issues in design and testing of welds are shown schematically. (QA/QC is Quality Assurance, Quality Control and SCC is Stress Corrosion Cracking)

It is necessary at the outset to clearly fix the relationships of computer methods to experimental investigation, Figure 1-3.

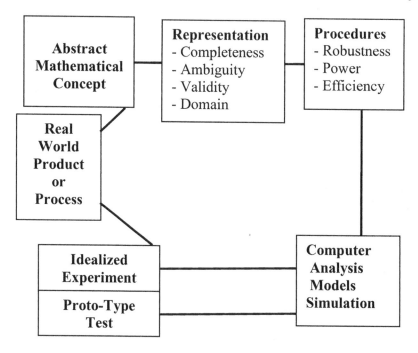

Figure 1-3: The relationships between the real world, experiments, mathematical abstraction and computer analysis.

Experiments tend to fall into two broad categories. Some are based on clearly understood theory where a strong attempt is made to exclude extraneous factors. Measurement of Young's modulus or thermal conductivity of a particular alloy fall into this category. On the other hand, when experiments deal with complex phenomena that do not have a clearly understood theory, a strong attempt is made to include any factor that may be relevant. Developing a narrow gap welding process is an example of this second category.

1.2 Brief History of Computational Welding Mechanics

Historically, arc welding began shortly after electrical power became available in the late *1800s*. Serious scientific studies date from at least the *1930s*. The failure of welded bridges in Europe in the

1930s and the American Liberty ships in World War *II* did much to stimulate research in welding in the *1940s*.

In the *USA*, the greatest attention was focused on developing fracture mechanics and fracture toughness tests. This could be interpreted as a belief that welding was too complex to analyze and therefore they chose an experimental approach that relied heavily on metallurgical and fracture toughness tests. The steady state heat transfer analysis of Rosenthal was an exception [3].

Russia took a different approach. The books of Okerblom [4] and Vinokurov [5] are a rich record of the analysis of welded structures including multi-pass welds and complex structures.

Over time the main techniques for solving heat transfer problems were changing with growing computer capacity. The strategy for analyzing welds and numerical methods (finite difference and finite element analysis) began in the *1960s* with the pioneering work of Hibbitt and Marcal [6], Friedman [7], Westby [2], Masubuchi [8] and Andersson [9]. Marcal [11] made an early summary of experiences from welding simulation. Chihoski sought a theory to explain why welds cracked under certain conditions but not others [51, 52 and 53]. Basically, he imaged that the weld was divided into longitudinal strips and into transverse strips. He then computed the thermal expansion and contraction in each of these strips due to the temperature field of edge and butt welds. He concluded that a small intense biaxial compression stress field exists near the weld pool. Ahead of the compression field there may be a gap or a tensile stress field. Behind the compression field a tensile field or a crack can appear. The startling aspect of Chihoski's theory was that by varying the welding procedure, the position of this compressive field could be controlled. To support this theory, Chihoski developed a Moire'fringing technique to measure displacements during edge and butt weld. Chihoski used his theory to understand and solve a number of common problems: cracks and microcracks, forward gapping, upset and part distortion, sudden changes in current demand and unexpected responses to welding gaps. He considered the position and pressure of hold down fingers; the influence of localized heating or cooling; and the effects of gaps. He argued that these parameters could be optimized to obtain crack free welds.

These authors consider Chihoski's papers to be among the most important in computational welding mechanics because he combined experience in production welds with an insight into weld mechanics that enabled him to conceive a theory that rationalized his observations and predicted solutions to his problems.

The reviews by Karlsson [12 and 13], Smith [17], Radaj [18 and 19] and Goldak [1, 14, 15, and 16] include references to simulations performed up to 1992. The research in Japan is reviewed by Ueda [20 and 21] and Yurioka and Koseki [22]. But Finite Element Analysis *(FEA)* methods gained a wide acceptance only over the last decade [10, 23, 24, 25, and 46].

The Moire'fringing technique of Chihoski is today one of the most powerful means of assessing *FEA* of stress and strain in welds (also see Johnson [55] for more on Moire' fringing methods for measuring strain in welds.).

The use of finite difference methods is more a transition between analytical and finite element methods. The main advantage of the finite difference method is that it is rather simple and easily understandable physically [10 and 26].

The finite element method has achieved considerable progress and powerful techniques of solving thermal-mechanical manufacturing process such as welding [1, 27, 28, 29 and 30]. Runnemalm presents in a dissertation thesis [24] the development of methods, methodologies and tools for efficient finite element modeling and simulation of welding. The recently published dissertation of Pilipenko [10] presents the development of an experimental, numerical and analytical approach to the analysis of weldability.

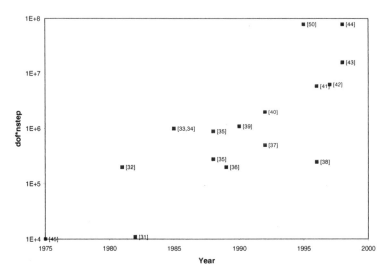

Figure 1-4: Size of computational models of welding measured by degree of freedom multiplied by number of time steps versus year of publication of work [47 and 49].

Lindgren review's [47, 48 and 49] summarized the direction for future research, which consists of three parts. Part 1 [47] shows that increased complexity of the models gives a better description of the engineering applications. The important development of material modeling and computational efficiency are outlined in part 2 and 3 [48 and 49], respectively. Figure 1-4 shows the increase in the size of the computational models in welding simulation during the last decades.

1.3 Mechanical Behavior of Welds

From the industry point of view, the most critical mechanical effects of welding are cracking, distortion and buckling. The influence of welds on crack propagation in stress corrosion cracking, fatigue and fracture is also a concern.

The difficulties of obtaining relevant material properties, in making experimental measurements to validate predictions and the complexity of the mathematical descriptions have all inhibited progress.

In practice however there are many aspects of welding that have made the rigorous analysis of welds a challenging procedure. At the macroscopic level a weld can be considered to be a thermo-mechanical problem of computing transient temperature, displacement, stress and strain. At the microscopic level, it can be considered to be a metal physics problem of computing the phase transformations including grain growth, dissolution and precipitation.

In addition to the capability of computing the microstructure and fracture mechanics, this will require a model for hydrogen diffusion along dislocations and grain boundaries as well as volume diffusion. Hydrogen diffusion can be a function of stress. It will require a model of hydrogen trapping, including a model for inclusions [15].

The mechanical behavior of welds is sensitive to the close coupling between heat transfer, microstructure evolution and thermal stress analysis. Figure 1-5 describes the coupling between the different fields in the modeling of welding. Although the effects of microstructure and stress-strain evolution on heat transfer are not large, the effect of temperature on the microstructure and thermal stress is dominant. In addition, the coupling between microstructure and thermal stress can be strong and subtle (the dominant couplings in welding are shown with bold lines, the secondary couplings are shown with dotted lines, Figure 1-5). Phase transformations, can dominate the stress analysis, [6 and 15].

The modeling of the fluid flow is not included in Figure 1-5, because the effect of the fluid flow on the deformation and stress field can be considered as negligible [24]. However, if geometrical changes close to the weld pool are of primary interest, modeling the fluid flow will be essential.

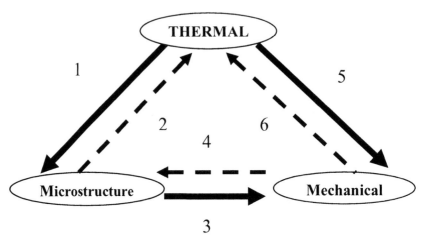

Coupling Explanation

1 Transformation Rates (microstructure evolution depends on temperature).

2 Latent heats (each phase transformation can have an associated latent heat). They act as a heat sink on heating and as a heat source on cooling.

3 Phase Transformations (volume changes due to phase changes, plastic and elastic material behavior depend on the microstructure of material).

4 Transformation Rates (microstructure evolution can particularly martensitic and bainitic transformations, depend on mechanical deformation).

5 Thermal Expansion (mechanical deformations depend on temperature).

6 Plastic Work (mechanical deformation generates heat in the material and changes the thermal boundary conditions). In most welding processes this effect is very small.

Figure 1-5: Coupling between different fields in welding analysis.

In the past ten years considerable progress has been made in developing numerical methods to solve this coupled problem with increasing speed and accuracy.

Realistic welds may involve numerous passes, each of which contributes to the mechanical and metallurgical effects. Interactions between thermal, mechanical, metallurgical, and in the molten pool chemical and fluid processes are complex.

1.4 Major objective of this book

The content of this book is divided into nine chapters. This short chapter is an introduction to the following chapters that discuss various subjects related to "Computational Welding Mechanics". "Computer Simulation of Welding Processes" is the title of chapter two. The third chapter, "Thermal Analysis of Welds" provides the necessary background information for thermal stress fields, structure distortion and deformation in order to enable those readers whose knowledge in these areas is limited, to understand the remainder of book. The heat supplied by the welding process produces complex thermal cycles in the weldment and these in turn cause changes in the microstructure of the heat affected zone, as well as the transient thermal stress. This results in the creation of residual stresses and distortions in the finished product. In order to analyze these problems chapters four to six discuss analyzing the heat flow during welding, microstructure evolution and distortion analysis. Applications of welding in industrial fields and the feasibility of real-time analysis will be presented in chapters seven. Cracks may occur in the weld metal or in the heat affected zone. Chapter eight discusses the fracture of the weldment. The authors' preferred input data for a Computational Weld Mechanics analysis are described in chapter nine.

References

1. Goldak J.A., Oddy A., Gu M., Ma W., Mashaie A and Hughes E. Coupling heat transfer, microstructure evolution and thermal stress analysis in weld mechanics, IUTAM Symposium, Mechanical Effects of Welding, June 10-14, Lulea Sweden, 1991
2. Westby O. Temperature distribution in the workpiece by welding, Department of Metallurgy and Metals Working, The Technical University, Trondheim Norway, 1968

3. Rosenthal, D. The theory of moving sources of heat and its application to metal treatments, Trans ASME, 1946, vol. 68, p 849-865

4. Okerblom, N.O. The calculations of deformations of welded metal structures, London, Her Majesty's Stationary Office, 1958

5. Vinokurov, V. A. Welding stresses and distortion, The British Library Board, 1977

6. Hibbitt H.D., Marcal P.V. A numerical thermo-mechanical model for the welding and subsequent loading of a fabricated structure, Comp. & Struct., Vol. 3, pp 1145-1174, 1973.

7. Friedman, E. Analysis of weld puddle distortion, Welding J Research Suppl. 1978, pp 161s-166s

8. Masubuchi. K. Analysis of welded structures, Pergamon Press, 1980

9. Andersson, BAB. Thermal stresses in a submerged-arc welded joint considering phase transformations, J Eng. & Tech. Trans. ASME, Vol. 100, 1978, pp 356-362

10. Pilipenko A (2001) Computer simulation of residual stress and distortion of thick plates in multielectrode submerged arc welding. Doctoral thesis, Norwegian University of Science and Technology.

11. Marcal P. Weld problems, structural mechanics programs, Charlottesville, University Press, pp 191-206, 1974

12. Karlsson L. Thermal stresses in welding, in R.B. Hetnarski (ed.), Thermal Stresses, Vol. I, Elsevier Science Publishers, pp 300, 1986

13. Karlsson L. Thermo mechanical finite element models for calculation of residual stresses due to welding, in Hauk et al (eds), Residual Stresses, DGM Informationsgesellschaft Verlag, p 33, 1993.

14. Goldak J., Patel B., Bibby M. and Moore J Computational weld mechanics, AGARD Workshop-Structures Materials 61st Panel meeting, 1985

15. Goldak J. Modeling thermal stresses and distortions in welds, Proc. of the 2nd Int. Conf. on Trends in Welding Research, p 71, 1989

16. Goldak J., Bibby M., Downey D. and Gu M. Heat and fluid flow in welds. Advanced Joining Technologies, Proc. of the Int. Congress on Joining Research, p 69, 1990

17. Smith S.D., A review of weld modeling for the prediction of residual stresses and distortions due to fusion welding, Proc. of the fifth Int. Conf. on Computer Technology in Welding. Paper 4, 1992.

18. Radaj D. Heat effects of welding: temperature field, residual stress, distortion. Springer, 1992

19. Radaj D. Finite element analysis of welding residual stresses, Proc. of 2nd Int. Conf. on Residual Stresses , p 510, 1988

20. Ueda Y. and Murakawa H. Applications of computer and numerical analysis techniques in welding research, JWRI, Vol. 13, No. 2, pp 165-174, 1984

21. Ueda Y., Murakawa H., Nakacho H. and Ma N-X., : Establishment of computational weld mechanic, JWRI, vol. 24, pp 73-86.1995

22. Yurioka N. and Koseki T., Modeling activities in Japan, in Cerjak H.(ed.), Mathematical of Modeling of Weld Phenomena 3, The Institute of Materials, pp 489, 1997. Also in JWRI, vol. 25, pp 33-52,1996
23. Moltubak T. Strength mismatch effect on the cleavage fracture toughness of the heat affected zone of steel welds, Doctoral Thesis, Trondheim, 1999
24. Runnemalm H (1999). Efficient finite element modeling and simulation of welding. Doctoral Thesis, Lulea.
25. Wikander L (1996). Efficient thermo-mechanical modeling of welding. Doctoral thesis, Lulea.
26. Pilipenko A.Yu. (1997). Analysis of the temperature distribution during GTA welding of thick-walled pipes. Master Thesis, St.-Petersburg Technical University.
27. Goldak J., Breiguine V., Dai N., Zhou J. Thermal stress analysis in welds for hot cracking, AMSE, Pressure Vessels and Piping Division PVP, Proceeding of the 1996 ASME PVP Conf., July 21-26, Montreal
28. Gundersen O. Mathematical modeling of welding. A state of the art review, SINTEF Material Technology, Trondheim 1997
29. Lindgren L-E (1992). Deformation and stresses in butt-welding of plates. Doctoral Thesis, Lulea, Sweden.
30. ABAQUS Version 5.6, ABAQUS user's manual, Hibbit, Karlsson & Sorensen, Inc. 1997, Providence, Rhode Island, USA
31. Argyris J.H., Szimmat J. and Willam K.J. Computational aspects of welding stress analysis, Computer Methods in Applied Mechanics and Engineering, vol. 33 pp 635-666, 1982
32. Andersson B. and Karlsson L. Thermal stresses in large butt-welded plates, J. of Thermal Stresses, vol. 4, pp 491-500, 1981
33. Josefson B.L., Residual stresses and cracking susceptibility in butt-welded stainless steel pipes, 4th Int. Conf. On Numerical Methods in Thermal Problems, p 1152, 1985
34. Jonsson M., Karlsson L. and Lindgren L.E., Deformations and stresses in butt-welding of large plates, in R.W.Lewis (ed.), Numerical Methods in Heat Transfer, Vol. III, p 35, Wiley 1985
35. Lindgren L.E. and Karlsson L. Deformation and stresses in welding of shell structures, Int. J. for Numerical Methods in Engineering, Vol. 25, pp 635-655, 1988
36. Karlsson C.T. Finite element analysis of temperatures and stresses in a single-pass butt-welded pipe influence of mesh density and material modeling, Eng. Comput., Vol. 6, pp 133-142, 1989
37. Oddy AS, Goldak JA and McDill JMJ. Transformation plasticity and residual stresses in single-pass repair welds, ASME J Pressure Vessel Technology, Vol. 114, pp 33-38, 1992
38. Ravichandran G., Raghupathy V.P., Ganesan N. and Kirshnakumer R. Analysis of transient longitudinal distortion in fillet welded T-beam using FEM with degenerated shell element, Int. J. for Joining of Materials, Vol.8, No. 4, pp 170-179, 1996

39. Karlsson R.I. and Josefson B.L. Three-dimensional finite element Analysis of temperatures and stresses in a single-pass butt-welded pipe, ASME Pressure Vessel Technology, Vol. 112, pp 76-84, 1990
40. Goldak J., Oddy A. and Dorling D. Finite element analysis of welding on fluid filled pressurized pipelines, Proc. Of the 3rd Int. Conf. On Trends in Welding Research, p 45, 1992
41. Oddy AS and McDill JMJ. Residual stresses in weaved repair welds, Proc. Of 1996 ASME Pressure Vessel & Piping Conf., p 147, 1996
42. Lindgren L-E., Haeggblad H-A., McDill JMJ. and Oddy AS. Automatic remeshing for three-dimensional finite element simulation of welding, Computer Methods in Applied Mechanics and Engineering, Vol. 147, pp 401-409, 1997
43. Oddy AS, McDill JMJ, Braid JEM., Root JH. and Marsiglio F. Predicting residual stresses in weaved repair welds, Proc. Of the 5th Int. Conf. On Trends in Welding Research, p 931, 1998
44. Fricke S., Keim E. and Schmidt J. Modeling of root formation during the welding process with the help of the 3d finite element method, In H. Cerjak (ed.), Mathematical Modeling of Weld Phenomena 4, The Institute of Materials, p 649, 1998
45. Friedman E. Thermo-mechanical analysis of the welding process using the finite element method, ASME J. Pressure Vessel Technology, Vol. 97, No. 3, pp 206-213, Aug. 1975
46. Volden L. (1999). Residual stress in steel weldments. Experiments, mechanism and modeling. Doctoral Thesis, Trondheim.
47. Lindgren L-E. Finite element modeling and simulation of welding Part I Increased complexity, J of Thermal Stresses 24, pp 141-192, 2001
48. Lindgren L-E. Finite element modeling and simulation of welding Part II Improved material modeling, J of Thermal Stresses 24, pp 195-231, 2001
49. Lindgren L-E. Finite element modeling and simulation of Welding Part III, Efficiency and integration, J of Thermal Stresses 24, pp 305-334, 2001
50. Cowles J.H., Blanford M., Giamei A.F. and Bruskotter: Application of three dimensional finite element analysis to electron beam welding of a high pressure drum rotor, proc. of the 7th Int. Conf. Modeling of Casting, Welding and Advanced Solidification processes. The Minerals Metals & Materials Society. 347, 1995.
51. Chihoski Russel A. Understanding weld cracking in Aluminum sheet, Welding Journal, Vol. 25, pp 24-30, Jan. 1972
52. Chihoski Russel A. The character of stress fields around a weld arc moving on Aluminum sheet, Welding Research Supplement, pp 9s-18s, Jan. 1972
53. Chihoski Russel A. Expansion and stress around Aluminum weld puddles, Welding Research Supplement, pp 263s-276s, Sep. 1979
54. Oddy AS, McDill JMJ and Goldak JA. A general transformation plasticity relation for 3D finite element analysis of welds, Euro J Mech., Vol. 9, No. 3, pp 253-263, 1990

55. Johnson L. Formation of plastic strains during welding of Aluminum alloys, Welding Research Supplement, pp 298s-305s, July 1973

Chapter II

Computer Simulation of Welding Processes

2.1 Introduction and Synopsis

Computer aided Design *(CAD)* and analysis and Computer Integrated Manufacturing *(CIM)* are popular topics for research. Computer aided drafting and engineering analysis such as Finite Element Method *(FEM)* are well developed. Finger & Dixon [1 and 2] categorize research areas in their review of theory and methodology of mechanical design. They agree that computer-based models provide the best opportunity to improve theories of design and increase their acceptance. Simulations are useful in designing the manufacturing process as well as the manufactured component itself.

Figure 2-1 shows an overview of some of the activities that will be performed during the product development cycle adopted from [3 and 4].

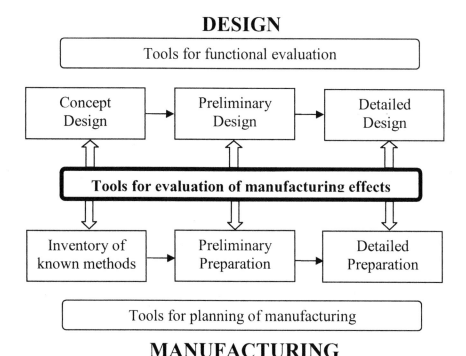

DESIGN

Tools for functional evaluation

| Concept Design | Preliminary Design | Detailed Design |

Tools for evaluation of manufacturing effects

| Inventory of known methods | Preliminary Preparation | Detailed Preparation |

Tools for planning of manufacturing

MANUFACTURING

Figure 2-1: Tools for evaluation of manufacturing effects between functional evaluation and planning of manufacturing (vertical) also between concept and details (horizontal), from [3 and 4].

In Figure 2-1, the upper part presents the traditional tools for functional evaluation and the lower part show the systems for computer aided design and finite element programs. Design engineers will use these tools, especially to improve the product and its functionality as much as possible. Weld simulation belongs to the middle part which includes tools for evaluation of manufacturing effects. This covers the most negative effects of manufacturing, on the properties in different phases of the process leading to defects in material state, form accuracy, measurement tolerance, strength, hardness and other quality features. The development of tools dedicated to supporting the evaluation of manufacturing effects has lagged due to the complexity of simulation processes. In Figure 2-1 the lower part demonstrates the current tools for planning of manufacturing, and above it, systems for programming robots and controlling material flow. Manufacturing engineers will use these tools especially to improve the production sequence and flow of material in the factory increasing the revenue of the manufactured product. Development steps from concept to detail in design and in manufacturing are shown in boxes. These steps must be run concurrently (simultaneous engineering), for which computing tools are inevitable requirements. Effort will be made in the entire process range, from raw material to the finished product, to include computers and simulations.

The calculation of weld simulation could not ignore computing tools in the future for innovative development of weld processes, weld design and their materials. Chihoski's [5] recommendation still says it best: "A changed set of conditions often changes the weld quality too subtly to be seen, except in large quantities, and there are too many possible changes to try. Hit or miss changes in the perfect lab (the production shop) are often not permitted. It would seem then to be of great use to the welding industry to develop and evolve computer programs that rigorously portray the stress and strain arrangements for different weld conditions. This route may be the only path from the current state of technology to the ideal in scientific promise, where a manufacturer who chooses an alloy and thickness and weld conditions can compute the value of each of the other weld conditions that minimize production problems".

2.2 The Computing Environment

2.2.1 Computational Geometry

In computer aided drafting, the computer stores the lines and symbols from the drawing, but does not understand them and cannot interpret them. With wire frame models that store vertices and edges, the computer has some understanding of the geometry and can rotate an object or at least its wire frame representation. However, the ambiguity of wire frames prevents the program from computing some values such as the volume of an object.

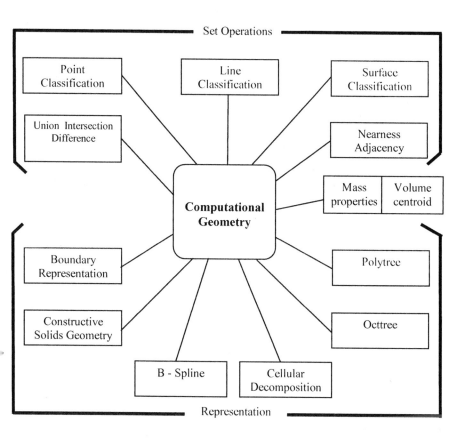

Figure 2-2: The representations and set operations that have been developed for computational geometry are shown above.

Modern computational geometry has developed representations and operations that overcome these difficulties, Figure 2-2. The best known of these is solids modeling which was developed to provide a mathematically complete representation of the geometry of an object [6].

Mathematicians usually approach geometry from one of two ways. Algebraic topology considers volume in *3D* space, e. g., a cube. The boundary of the volume is an oriented surface or set of surfaces of zero thickness. The boundary a surface is a curve. The boundary of a curve is a point. Point set topology considers *3D* space to filled with an infinite set of points. Each point is either inside, outside or on the boundary of a *3D* object. An *FEM* mesh is best treated from the viewpoint of algebraic topology. In *FEM* analysis of temperature or stress, it is usual to take the viewpoint of point set topology.

Modeling is the process of preparing a computational model. To start creating the analytical model we first have to decide what kind of geometric approximation would be most suitable for our purpose. Figure 2-3 shows several alternatives of geometry.

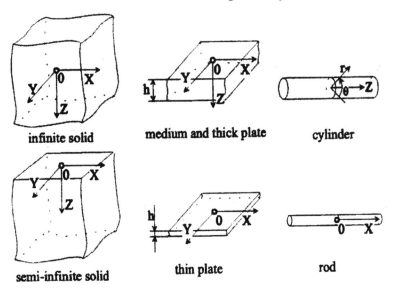

Figure 2-3: Different schemes of idealized body configuration, adopted from [18].

The next step towards analytical model creation is to choose the weld related approximation. Pilipenko [18] suggests the investigated problem can be classified by: time of action, mobility and dimensions. The time of action can be instantaneous or continuous and mobility, stationary or mobile. By the area of distribution, it can be presented as a point (dim=0), line (dim=1), plane (dim=2) and volume (dim=3).

To get a general view, all the classifications are presented in the scheme on Figure 2-4.

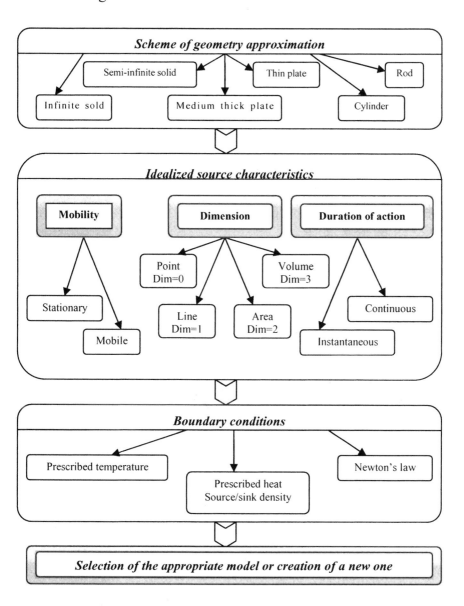

Figure 2-4: Flow chart of the model selection process, from [18].

At present many different analytical models are known. Lindgren presents a review of several in [22]. The earliest numerical predictions of residual stresses were probably those of Tall [21], Vinokurov [76] and Okerblom [75]. Tsuji [23] performed similar calculations. The mechanical analysis was essentially one-dimensional, although the analytic solution for the temperatures was two dimensional in the work by Tall. The earliest two-dimensional finite element analysis appeared in the early *1970s*, by Iwaki and Masubuchi [24], Ueda and Yamakawa [25, 26 and 27], Fujita et al. [28], Hibbitt and Marcal [29] and Friedman [30]. The early simulations by Fujita et al. [31] and Fujita and Nomoto [32] used only a thermo elastic material model. Although in welding the reality is *3D* and transient, and *3D* analysis is the most general formulation, to minimize computing costs many analysts have used lower dimensional models. However, they have long sought to perform full *3D* analysis of welds. The first analyses of this kind were reported in *1986* [33-35].

2.2.2 Models for Welding Heat Sources

Theoretical Formulations

The basic theory of heat flow that was developed by Fourier and applied to moving heat sources by Rosenthal [38] and Rykalin [37] in the late *1930s* is still the most popular analytical method for calculating the thermal history of welds. As many researchers have shown, Rosenthal's point or line heat source models are subject to a serious error for temperatures in or near the fusion zone *(FZ)* and heat affected zone *(HAZ)*. The infinite temperature at the heat source assumed in this model and the temperature sensitivity of the material thermal properties increases the error as the heat source is approached. The effect of these assumptions and others on the accuracy of temperature distributions from the Rosenthal analysis has been discussed in detail by Myers et al. [39].

To overcome most of these limitations several authors have used the finite element method *(FEM)* to analyze heat flow in welds. Since Rosenthal's point or line models assume that the flux and temperature is infinite at the source, the temperature distribution has many similarities to the stress distribution around the crack tip in linear elastic fracture mechanics. Therefore many of the *FEM* techniques developed for fracture mechanics can be adapted to the Rosenthal model. Certainly it would be possible to use singular *FEM* elements to analyze Rosenthal's formulation for arbitrary geometries. This would retain most of the limitations of Rosenthal's analysis but would permit complex geometries to be analyzed easily. However, since it would not account for the actual distribution of the heat in the arc and hence would not accurately predict temperatures near the arc, this approach is not pursued here. Pavelic et al. [40] first suggested that the heat source should be distributed and he proposed a Gaussian distribution of flux deposited on the surface of the workpiece. Figure 2-5 represents a circle surface heat source and a hemispherical volume source, both with Gaussian normal distribution (bell shape curves), in a mid-thick plate. The geometrical parameters of heat flux distribution are estimated from the results of weld experiments (molten zone, size and shape and also temperature cycle close to molten zone).

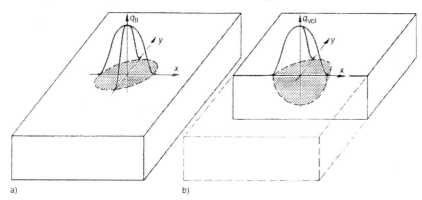

Figure 2-5 Heat source distribution in weldment: circular form surface source (a) and hemispherical volume source (b); Gaussian distribution of the surface related source density q_{fl} and volume related source density q_{vol} [3].

The subsequent works of Andersson [41], Krutz and Segerlind [42] and Friedman [43] are particularly notable. In References [42] and [43] Pavelic's 'disc' model is combined with *FEM* analysis to achieve significantly better temperature distributions in the fusion and heat affected zones than those computed with the Rosenthal model.

While Pavelic's 'disc' model is certainly a significant step forward, some authors have suggested that the heat should be distributed throughout the molten zone to reflect more accurately the digging action of the arc. This approach was followed by Paley [44] and Westby [45] who used a constant power density distribution in the fusion zone (FZ) with a finite difference analysis, but no criteria for estimating the length of the molten pool was offered. In addition, it is difficult to accommodate the complex geometry of real weld pools with the finite difference method.

The analyst requires a heat source model that accurately predicts the temperature field in the weldment. A non-axisymmetric three-dimensional heat source model which is proposed in this book achieves
this goal. It is argued on the basis of molten zone observations that this is a more realistic model and more flexible than any other model yet proposed for weld heat sources. Both shallow and deep penetration welds can be accommodated as well as asymmetrical situations.

The proposed three-dimensional 'double ellipsoid' configuration heat source model is the most popular form of this class of heat source models [46]. It is shown that the 'disc' of Pavelic et al [40] and the volume source of Paley and Hibbert [44] and Westby [45] are special cases of this model. In order to present and justify the double ellipsoid model, a brief description of the Pavelic 'disc' and of the Friedman [43] modification for *FEM* analysis is necessary. In addition, the mathematics of the disc is extended to spherical, ellipsoidal and finally to the double ellipsoidal configuration. Finally, it is pointed out the most general form of this class of model has a general heat source distribution function.

Model Considerations

The interaction of a heat source (arc, electron beam or laser) with a weld pool is a complex physical phenomenon that still cannot be modeled rigorously. It is known that the distribution of pressure and shear from the arc source, droplets from the electrode the effects of surface tension, buoyancy forces and molten metal viscosity combine to cause weld puddle distortion and considerable stirring. Because of the arc 'digging' and stirring, it is clear that the heat input is effectively distributed throughout a volume in the workpiece.

The 'disc' model is more realistic than the point source because it distributes the heat input over a source area. In fact, for a preheat torch that causes no melting this may be a very accurate model indeed. However, in the absence of modeling the weld pool free boundary position, the applied tractions, and convective and radiative conditions between the weld pool and the arc, some form of idealization of the heat source is necessary to achieve as good an approximate solution as one can afford. The disc model does not account for the rapid transfer of heat throughout the fusion zone *FZ*. In particular, it is not possible to predict the deep penetration *FZ* of an electron beam *EB* or laser weld with the surface disc model. A comparison of calculated thermal history data (disc model) with measured values during author's investigations [46] underscored the need for an 'effective volume source' such as the one suggested by Paley and Hibbert [44]. In addition, it was found necessary to generate a volume source with considerable flexibility, i.e., the double ellipsoid model. With less general shapes such as a hemisphere or a single ellipsoid significant discrepancies between the computed and measured temperature distributions could not be resolved.

The size and shape of the 'double ellipsoid' i.e., the semi-axes lengths, can be fixed by recognizing that the solid-liquid interface is the melting point isotherm. In reality the melting point is a function of curvature and the speed of the liquid-solid interface but the changes have been ignored in most models published to date. At the same time weld pool temperature measurements have shown that the peak temperature in the weld pool is often *300* to *500 °C* above the melting point. The accuracy with which the heat source model predicts the size and shape of the *FZ* and the peak temperatures is probably the most stringent test of the performance of the model. In the author's investigation [46] it was found that the most accuracy was obtained when the ellipsoid size and shape were equal to that of the weld pool. The non-dimensional system suggested by Christensen [48] can be used to estimate the ellipsoid parameters. Furthermore a Gaussian distribution is assumed centered at the origin of the heat source.

Gaussian Surface Flux Distribution

In the 'circular disc' model proposed by Pavelic et al [40], the thermal flux has a Gaussian or normal distribution in the plane, Figure 2-6:

$$q(r) = q(0)e^{-Cr^2} \tag{2-1}$$

where:

$q(r)$ = Surface flux at radius $r (W/m^2)$

$q(0)$ = maximum flux at the center of the heat source (W/m^2)

C = distribution width coefficient (m^{-2})

r = radial distance from the center of the heat source (m)

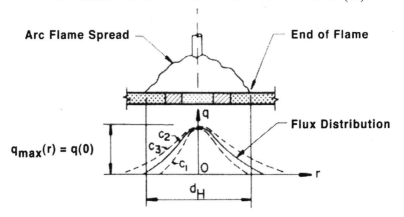

Figure 2-6: Circular disc heat source [40]

A simple physical meaning can be associated with C. If a uniform flux of magnitude $q(0)$ is distributed in a circular disk of diameter $d = 2/\sqrt{C}$, the rate of energy input would be ηIV, i.e., the circle would receive all of the energy directly from the arc. Therefore the coefficient, *C*, is related to the source width; a more concentrated source would have a smaller diameter *d* and a larger value of *C*. To translate these concepts into practice, the process model for the normal distribution circular surface of the heat flux for laser beam welding and submerged arc welding, is illustrated in Figure 2-7.

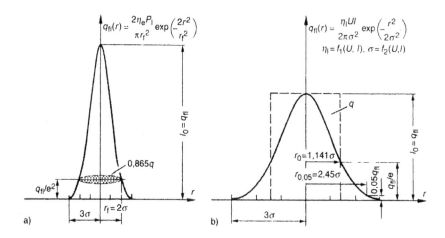

Figure 2-7: Normal distribution circular surface heat source and related parameters (radial distance from center σ) by laser beam welding (a) and by submerged arc welding (b); with laser efficiency coefficient η_e, the laser power P_1, the focus diameter $2r_f$, the arc efficiency coefficient η_1, voltage of arc V, the current I, the maximum flux at the center of the heat source q(0) equal intensity I_0, Euler number e=2,71828...; from Sudnik (unpublished) and Radaj [3].

The curve of the submerged arc welding is wide and low and the curve of the laser beam welding is in contrast, narrow and high. The laser beam has high intensity and a small diameter. The laser power *P_l* on the surface of the workpiece considering the efficiency coefficient is used as heat flux (power) *q* and/or heat power density q_{fl}. The circular normal distribution is described by the focus radius $r_f = 2\sigma$ which contains *86%* of the heat power.

The arc welding has less intensity and larger diameter. The arc power *VI* on the surface of the workpiece considering the arc efficiency coefficient is used as heat flux *q* and/or heat power density q_{fl}. For the circular normal distribution two different descriptions are used. The source radius $r_{0.05}$, in this heat power density is reduced by 5% (Rykalin [37] or the source radius r_0 of a same power source with constant heat power density (Ohji et al [51, 73]). The particular problem which exists for the definition of the heat source in arc welding is that both the arc efficiency η_I and the radial distance from the center σ are functions of the voltage V and the current I of the arc (Sudnik and Erofeew [72]).

The effective radius r_0 and/or $r_{0.05}$ of the circular surface heat source are derived from the maximum surface width of the molten zone. The heat flow density from the surface $q_{fl}(x,y)$ for known r_0 and/or $r_{0.05}$ is defined from the power data of the weld heat source, considering the heat transfer efficiency. This is shown in Figure 2-7 for the laser beam welding and arc welding. If weld metal is added the related heat should be considered.

Experiments have shown that a significant amount of heat is transferred by radiation and convection from the arc directly to the solid metal without passing through the molten pool. Based on this observation, Pavelic et al. [40] developed a correlation showing the amount and the distribution of this heat over the solid material. In their study, provisions were made for convective and radiative losses from the heated plate to the surroundings as well as variable material properties.

Friedman [43] and Krutz and Segerland [42] suggested an alternative form for the Pavelic 'disc'. Expressed in a coordinate system that moves with the heat source as shown in Figure 2-8, Eq. (2-2) takes the form:

$$q(x,\xi) = \frac{3Q}{\pi c^2} e^{-3x^2/c^2} e^{-3\xi^2/c^2} \qquad (2-2)$$

where:

Q = energy input rate *(W)*

c = is the characteristic radius of heat flux distribution *(m)*

It is convenient to introduce an (x, y, z) coordinate system fixed in the workpiece. In addition, a lag factor τ is needed to define the position of the source at time $t = 0$, Figure 2-8.

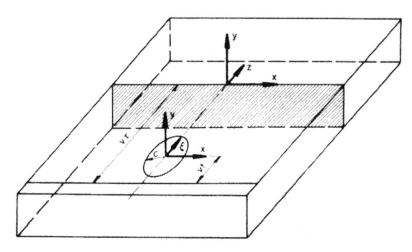

Figure 2-8: Coordinate system used for the FEM analysis of disc model according to Krutz and Segerlind [42]

The transformation relating the fixed *(x,y,z)* and moving coordinate system *(x,y,ξ)* is:

$$\xi = z + v(\tau - t) \tag{2-3}$$

where v is the welding speed (*m/s*). In the (x, y, z) coordinate system Eq. (2-2) takes the form:

$$q(x,z,t) = \frac{3Q}{\pi c^2} e^{-3x^2/c^2} e^{-3[z+v(\tau-t)]^2/c^2} \tag{2-4}$$

For $x^2 + \xi^2 < c^2$. For $x^2 + \xi^2 > c^2$, $q(x,\xi,t) = 0$

To avoid the cost of a full three-dimensional *FEM* analysis some authors assume negligible heat flow in the longitudinal direction; i.e., $\partial T/\partial z = 0$. Hence, heat flow is restricted to an *x-y* plane, usually positioned at $z = 0$. This has been shown to cause little error except for low speed high heat input welds [41]. The disc moves along the surface of the workpiece in the z direction and deposits heat on the reference plane as it crosses. The heat then diffuses outward (*x-y* direction) until the weld cools.

Hemi-spherical Power Density Distribution

For welding situations, where the effective depth of penetration is small, the surface heat source model of Pavelic, Friedman and Krutz has been quite successful. However, for high power density sources such as the laser or electron beam, it ignores the digging action of the arc that transports heat well below the surface of the weld pool. In such cases a hemispherical Gaussian distribution of power density (W/m^3) would be a step toward a more realistic model. The power density distribution for a hemispherical volume source can be written as:

$$q(x, y, \xi) = \frac{6\sqrt{3}Q}{c^3 \pi \sqrt{\pi}} e^{-3x^2/c^2} e^{-3y^2/c^2} e^{-3\xi^2/c^2} \qquad (2\text{-}5)$$

where $q(x, y, \xi)$ is the power density (W/m^3). Eq. (2-5) is a special case of the more general ellipsoidal formulation developed in the next section.

Though the hemispherical heat source is expected to model an arc weld better than a disc source, it, too, has limitations. The molten pool in many welds is often far from spherical. Also, a hemispherical source is not appropriate for welds that are not spherically symmetric such as a strip electrode, deep penetration electron beam, or laser beam welds. In order to relax these constraints, and make the formulation more accurate, an ellipsoidal volume source has been proposed.

Ellipsoidal Power Density Distribution

The Gaussian distribution of the power density in an ellipsoid with center at $(0, 0, 0)$ and semi-axis a, b, c parallel to coordinate axes x, y, ξ can be written as:

$$q(x, y, \xi) = q(0)e^{-Ax^2} e^{-By^2} e^{-C\xi^2} \qquad (2\text{-}6)$$

where $q(0)$ is the maximum value of the power density at the center of the ellipsoid.

Conservation of energy requires that:

$$2Q = 2\eta VI = 8\int_0^\xi \int_0^y \int_0^x q(0)e^{-Ax^2}e^{-By^2}e^{-C\xi^2}\,dx\,dy\,d\xi \qquad (2\text{-}7)$$

where;

η = Heat source efficiency

V = voltage

I = current

Evaluation of Eq. (2-7) produces the following:

$$2Q = \frac{q(0)\pi\sqrt{\pi}}{\sqrt{ABC}} \qquad (2\text{-}8)$$

$$q(0) = \frac{2Q\sqrt{ABC}}{\pi\sqrt{\pi}} \qquad (2\text{-}9)$$

To evaluate the constants, A, B, C, the semi-axes of the ellipsoid a, b, c in the directions x, y, ξ are defined such that the power density falls to $0.05\,q(0)$ at the surface of the ellipsoid. In the x direction:

$$q(a,0,0) = q(0)e^{-Aa^2} = 0.05\,q(0) \qquad (2\text{-}10)$$

Hence

$$A = \frac{\ln 20}{a^2} \approx \frac{3}{a^2} \qquad (2\text{-}11)$$

Similarly

$$B \approx \frac{3}{b^2} \qquad (2\text{-}12)$$

$$C \approx \frac{3}{c^2} \qquad (2\text{-}13)$$

Substituting A, B, C from Eqs. (2-11) to (2-13) and $q(0)$ from Eq. (2-9) into Eq. (2-6):

$$q(x,y,\xi) = \frac{6\sqrt{3}Q}{abc\,\pi\sqrt{\pi}}e^{-3x^2/a^2}e^{-3y^2/b^2}e^{-3\xi^2/c^2} \qquad (2\text{-}14)$$

The coordinate transformation, Eq. (2-3), Figure 2-8, can be substituted into Eq. (2-14) to provide an expression for the ellipsoid in the fixed coordinate system.

$$q(x,y,z,t) = \frac{6\sqrt{3}Q}{abc\,\pi\sqrt{\pi}}e^{-3x^2/a^2}e^{-3y^2/b^2}e^{-3[z+v(\tau-t)]^2/c^2} \qquad (2\text{-}15)$$

If heat flow in the *z* direction is neglected, an analysis can be performed on the *z-y* plane located at *z* = *0* which is similar to the 'disc' source. The power density is calculated for each time increment, where the ellipsoidal source intersects this plane.

Double Ellipsoidal Power Density Distribution

Calculation experience with the ellipsoidal heat source model revealed that the temperature gradient in front of the heat source was not as steep as expected and the gentler gradient at the trailing edge of the molten pool was steeper than experimental measurementse. To overcome this limitation, two ellipsoidal sources were combined as shown in Figure 2-9. The front half of the source is the quadrant of one ellipsoidal source, and the rear half is the quadrant of another ellipsoid. The power density distribution along the ξ axis is shown in Figure 2-9.

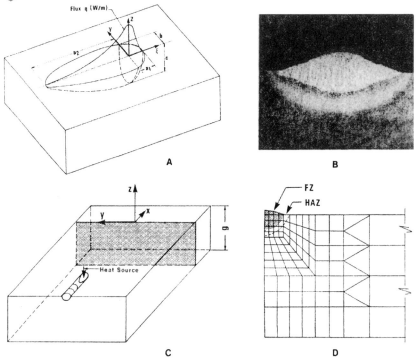

Figure 2-9: Double ellipsoid heat source configuration together with the power distribution function along the ξ axis (A), Cross-section of an SMAW weld bead

on a thick plate low carbon steel (V=30 volts, I=265 amps, v=3.8 mm/s, g=38 mm, T_0=20.5 °C), Reference plane concept and mesh used for the analysis will be discussed in the following chapters (D).

In this model, the fractions f_f and f_r of the heat deposited in the front and rear quadrants are needed, where $f_f + f_r = 2$. The power density distribution inside the front quadrant becomes:

$$q(x,y,z,t) = \frac{6\sqrt{3}f_f Q}{abc\pi\sqrt{\pi}} e^{-3x^2/a^2} e^{-3y^2/b^2} e^{-3[z+v(\tau-t)]^2/c^2} \qquad (2\text{-}16)$$

Similarly, for the rear quadrant of the source the power density distribution inside the ellipsoid becomes:

$$q(x,y,z,t) = \frac{6\sqrt{3}f_r Q}{abc\pi\sqrt{\pi}} e^{-3x^2/a^2} e^{-3y^2/b^2} e^{-3[z+v(\tau-t)]^2/c^2} \qquad (2\text{-}17)$$

In Eqs. (2-16) and (2-17), the parameters $a,\ b,\ c$ can have different values in the front and rear quadrants since they are independent. Indeed, in welding dissimilar metals, it may be necessary to use four octants, each with independent values of a, b and c

In cases where the fusion zone differs from an ellipsoidal shape, other models should be used for the flux and power density distribution. For example, in welds with a cross-section shaped as shown in Figure 2-10, four ellipsoid quadrants can be superimposed to more accurately model such welds.

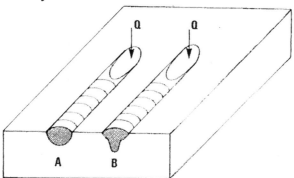

Figure 2-10: Cross-sectional weld shape of the fusion zone where a double ellipsoid is used to approximate the heat source (A), compound double ellipsoids must be superimposed to more accurately model such welds.

For deep penetration electron and laser beam welds, a conical distribution of power density which has a Gaussian distribution radially and a linear distribution axially has yielded more accurate results, Figure 2-11.

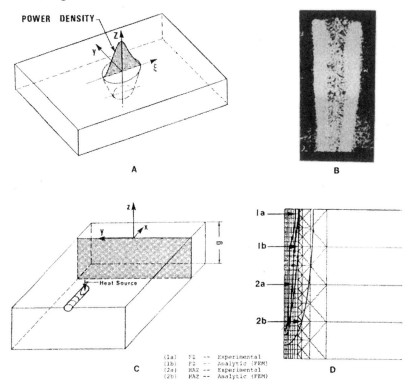

Figure 2-11: A conical weld heat source used for analyzing deep penetration electron beam or laser welds; conical source (A), typical electron beam weld (B), cross-sectional kinematic model with reference plane (C) and computed and measured Fusion zone (FZ) and heat affected zone (HAZ) boundaries, (V=70 kV, I=40 mA, v=4.23 mm/s, g=12.7 mm and T_0= 21°C).

The analyst must specify these functions or at least the parameters such as weld current, voltage, speed, arc efficiency and the size and position of the discs, ellipsoids and/or cones. In some cases the weld pool size and shape can be estimated from cross-sectional metallographic data and from weld pool surface ripple markings. If such data are not available, the method for estimating the weld pool dimensions suggested by Christensen [48] for arc welds and by Bibby et al [68] for deep penetration electron beam or laser welds should be used.

The size and shape of the heat source model is fixed by the ellipsoid parameters defined in Figure 2-10. Good agreement between actual and computed weld pool size is obtained if the size selected is about *10%* smaller than the actual weld pool size. If the ellipsoid semi-axes are too long then the peak temperature is too low and the fusion zone too small. The author's experience is that accurate results are obtained when the computed weld pool dimensions are slightly larger then the ellipsoid dimensions. This is easily achieved in a few iterations. Chakravarti et al [69] have studied the sensivity of the temperature field to the ellipsoid parameters.

On the one hand these distribution functions can be criticized as "fudge" factors. On the other hand, they do enable accurate temperature fields to be computed. Chosen wisely, varying any parameter changes the computed temperature field. It can be argued that they are needed to model the many complex effects that are quantitatively known, such as electrode angle, arc length, joint design and shielding gas composition [70].

2.2.3 Kinematic Models for Welding Heat Transfer

Having selected a model for the heat source, the analyst has the option of assuming that the heat flows only in cross-sectional planes, only in the plane of the plate, only in a radial direction or is free to flow in all three dimensions, Figure 2-12. Such assumptions are analogous to those applied to the displacement field in beams, plates and shells in structural analysis. Since assumptions restrict the orientation of the thermal gradient, it is suggested they be called kinematic models. Of course, these kinematic models are quite distinct from the heat source models described in the previous section. These three models are discussed in detail in [36].

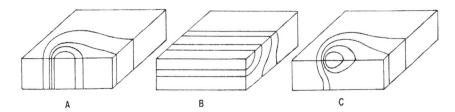

Figure 2-12: By constraining the thermal flux and hence the temperature gradients, the kinematic models are consistent with the temperature fields shown in; (A) in-plane, (B) cross-sectional models and (C) 3D models imply temperature.

The most popular has been the cross-sectional model which assumes that heat flow in the direction of welding is zero. This is surprisingly a subtle model. Strictly, it is limited to steady state analysis of prismatic geometries with the weld parallel to the prismatic axis. How can such a model possibly be accurate since some heat clearly must flow in the direction of welding? The answer is that the heat input has been modified to account for the heat that flows in the longitudinal direction. Because it uses relatively few *2D* elements, it is computationally one of the cheapest models.

The next most popular kinematic model has been the in-plane model. It assumes the heat flow normal to the weld plate is zero. The advantage of the in-plane over the cross-sectional model is its capability for computing the starting and ending transients. However, because it tends to require rather more elements and more time steps, the computational costs are higher [19]. It has not been used extensively.

In choosing a kinematic model, the analyst must balance accuracy against cost. In all cases, reality is three dimensional but the cost of analysis is the highest. Constraining heat flow to the plane of the plate can achieve useful accuracy for thin plates, particularly with deep penetration plasma, electron and laser beam welds. Assuming heat flows only in the cross-sectional plane can provide a useful and economical approximation for many welding situations. If the thermal diffusivity is sufficiently low and welding speed sufficiently high, this can be an accurate model. In particular, the results from a low cost cross-sectional analysis can be helpful in designing an efficient *3D* mesh.

The thermal shell model is a generalization of the in-plane model to curved surfaces. It assumes that no heat flows normal to the surface and that all heat flows in the tangent plane, i.e., the temperature gradient lies in the tangent plane. Where conditions hold to an adequate accuracy, the shell model has several advantages. The shell element usually has half the number of degrees of freedom and half the bandwidth, which reduces the computing costs. Even more important, it allows larger elements than the equivalent *3D* elements because they are better conditioned numerically.

In *3D* analyses the kinematic model is obviously correct. The issue is to reduce the computational cost to acceptable levels for meshes that achieve useful accuracy. The principal tools are optimizing the mesh and creating efficient solvers.

In *3D* analysis, brick elements are preferred because they tend to be more accurate and easier to use and interpret than tetrahedral elements. However, it is more difficult to grade a mesh with small brick elements near the welding arc to capture the rapid changes in temperature and with large brick elements far from the arc where the temperature varies slowly.

To solve these problems McDill et al. [10 and 47] developed a special brick element for graded *3D* meshes. The grading elements of McDill have made a major contribution in optimizing the mesh for the analysis of welds. These elements reduce the computational complexity from $O(n^7)$ for a cube with uniform mesh with n elements on an edge to $O(\log_2 n)$ for the cube with a graded mesh. In both cases, the problem has a point load on one corner of the cube and direct solvers are used. In this problem, there is no accuracy loss with the graded mesh. The advantages of this element are demonstrated by analyzing the problem shown in Figure 2-13. In comparison the cost and the memory requirements vary as $\log_2 n$ for the graded mesh. The larger the problem is the greater the advantage of the graded mesh. This clearly illustrates that the gains from better algorithms can easily outpace the gains from better computers.

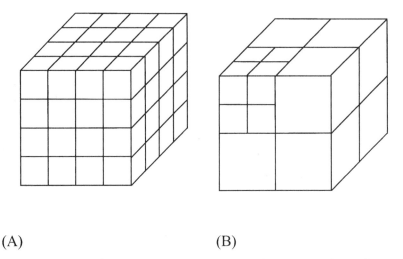

(A)   (B)

Figure 2-13: A uniform mesh n=4 (A) and McDill's grading scheme for a cube of n=4 (B). The number of elements is equal to n^3 and $7\log_2 n + 1$. The computational cost varies as n^7 and $\log_2 n$ (Where: n is the number of elements along an edge in a uniform mesh and, m is the number of recursive subdivisions of the upper left hand cube. Note: the size of the element in the upper left front corner is the same when $n=2^m$).

The second major weapon in the battle to reduce computing costs is the use of improved solvers. The use of incomplete Choleski conjugate gradient *(ICCG)* solvers with element-by-element preconditioning, which have been developed for fluids problems by Glowinski [12] and for structural problems by Jenning and Ajiz [13], has produced dramatic gains. These solvers do not factorize the global stiffness matrix completely.. Although the experience to date does not permit a definitive statement, Hughes [14 and 15] has developed an iterative method that promises to solve the problem and reduces *CPU* costs. The transient temperature fields shown in Figure 2-14 and 2-17 were computed on a workstation with such a solver.

The thinner the wall, the smaller is the three dimensional area around the heat source. An example is shown in Figure 2-14.

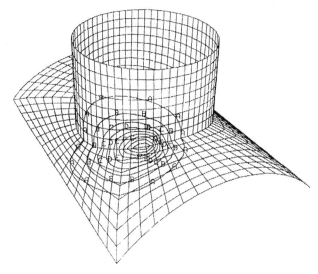

Figure 2-14: The contours of the transient temperature field are shown for a weld of a tee joint. The pipe is 40mm in diameter and the wall thickness is 2.5mm. Thermal shell elements are used to reduce the computing cost. The welding parameters are voltage 32, amperes 150, speed 1mm/s and power input 1.5 kJ/mm.

Since the *3D* model is needed near the arc and the shell model is most efficient far from the arc, these models can be combined to produce a more efficient model. Figure 2-15 shows such a composite *3D*-shell model. It requires a special transition element to join shell and *3D* brick elements.

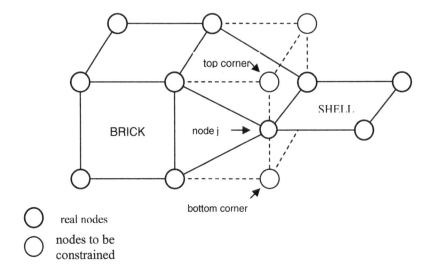

Figure 2-15: The transformation of a brick element to a transition element is shown. In particular, the relation between the transition node j and its top and bottom nodes is shown [67].

Gu in [67] presents a method to connect a three-dimensional element to the shell element. Figure 2-15 shows brick and shell elements and the transformation of a brick element to a transition element. The element is called a Shell-Brick transition element. It is not really a new element but a brick with constraints according to the shell assumption. It has more flexibility and can be added to other *FEM* packages without altering the original code.

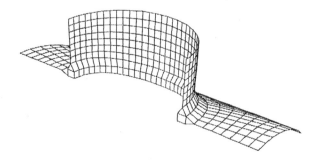

Figure 2-16: This tee joint is modeled with 4 noded thermal shell elements far from the weld joint, 8 noded 3D brick elements at the weld joint and special transition elements to join the shell and brick elements.

This composite kinematic model is more accurate than the shell mesh shown in Figure 2-16 and computationally cheaper than the mesh using *3D* bricks shown in Figure 2-17.

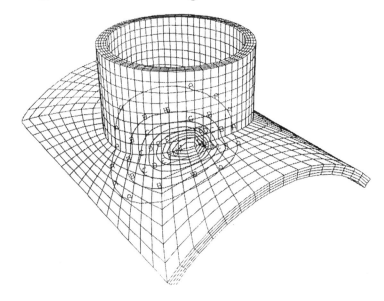

Figure 2-17: This tee joint is modeled with 8 noded 3D brick elements. This is the most accurate and the most expensive kinematic model.

The fundamental limitations of the most popular commercial mesh generators are that the mesh is difficult to change during an analysis; the mesh design requires expert judgment and it is labor intensive. Kela [7] and Sheperd and Law [8] developed a class of mesh generators which do not depend on expert judgment. They are fully automatic. Since as much as *90%* of the time and cost of a Finite Element Method *(FEM)* analysis of a complex structure can be absorbed by pre and post processing, progress in mesh generation is a critical factor that is pacing the development of computational weld mechanics. Automatic mesh generators open the possibility of changing the mesh at any, even every, time step. This is particularly important in welding where a fine mesh is needed near the molten pool and a coarse mesh far from the molten pool could decrease costs and improve performance by several orders of magnitude. The arbitrary Euler-Lagrange method [9] would allow a fine mesh that moves with the arc. The mesh far from the arc could be stationary. The mesh in between could move with a velocity that is interpolated.

The above examples suggest that rapid improvements in mesh generation methods that will reduce the cost of finite element analysis *(FEA)* of welds and permit real welding situations to be analyzed routinely. Many manufacturing processes are three-dimensional, nonlinear, transient processes in which the area of computational interest changes with time. Automatic finite element analysis AFEA with adaptive and dynamic mesh management has the potential to reduce substantially the cost of analysis for problems of this type.

In transient heat transfer analysis, the Finite Element Method *(FEM)* program must integrate a set of ordinary differential equations in time. Current commercial *FEM* programs use two point integration schemes that are either explicit or implicit. Explicit schemes use primarily element level operations. The cost per time step is small but time steps must be short because the element with the smallest critical time controls the time step. Implicit schemes solve a global set of linear equations. Each time step is more expensive but the time steps can be larger. Provided the load does not vary too rapidly in time, the longer time steps can be more efficient.

A recent development in integration schemes allows those elements with short critical time steps to be integrated implicitly with longer time steps and those elements with longer critical time steps to be integrated explicitly [17]. This can reduce computing costs significantly.

The generalization of this concept to use different time steps in different elements is called subcycling [16]. The mesh shown in Figure 2-18A is created by recursively subdividing the square m times.

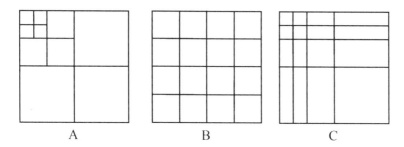

Figure 2-18: The mesh in (A) is obtained by recursively subdividing the square in the upper left hand corner. This can be significantly more efficient than the regular mesh shown in (B) or the grading strategy used in (C) for point like loads at the upper hand corner.

The number of elements is $1+3m$; the smallest element has edge length $L/2^m$ and critical time step $2^{2m}/\alpha L^2$. If each element is integrated only with its critical time step, the number of element time steps is $3(2^m - 1)$ or $O(2^m)$. If all elements are integrated with the smallest critical time step as is the current commercial practice, the number of element time steps is: $(1+3m)2^{2m}\alpha/L^2$ or $O(m2^{2m})$.

2.2.4 Evaluation of the Double Ellipsoid Model

In order to minimize the computing cost the initial analysis was done in the plane normal to the welding direction as shown in figures 2-19 and 2-20. Thus, heat flow in the welding direction was neglected. The above simplification is accurate in situations where comparatively little heat flows from the arc in the welding direction. This is reasonable when the arc speed is high. An estimate of the effect of this approximation has been given by Andersson [41] who argues that the errors introduced by neglecting heat flow in the direction of the moving electrode are not large, except in the immediate vicinity of the electrode.

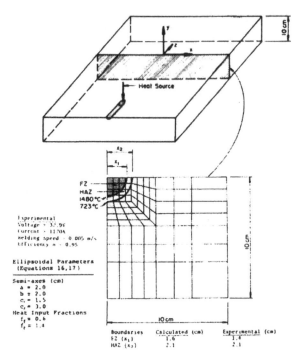

Figure 2-19: Experimental arrangement and FEM mesh for the thick section bead on plate weld [48 and 46].

In order to demonstrate the flexibility and assess the validity of the double ellipsoidal heat source model two quite different welding situations were considered.

The first case analyzed was a thick section *(10cm)* submerged arc bead-on-plate (low carbon structural steel) weld shown schematically in Figure 2-19. The welding conditions are also contained in the Figure.

Christensen [48] reported a *800* to *500* °C cooling time of *37* seconds for this weld and the *FZ* and HAZ sizes shown in the diagram. Shown also in the figure is the *FEM* mesh used to calculate these quantities. It is two-dimensional in x and y as previously explained. The temperature distribution in the 'cross-section analyzed' is calculated for a series of time steps as the heat source passes. In this way the *FZ* and *HAZ* cross-sectional sizes can be determined, and from the time step-temperature data the cooling time *800* to *500* °*C* is calculated.

The second welding situation is taken from the work of Chong [49], Figure 2-20.

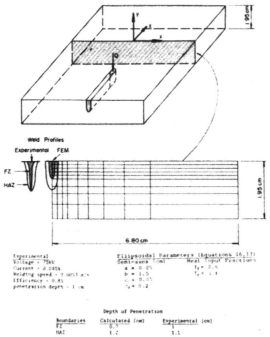

Figure 2-20: Experimental arrangement and FEM mesh for the deep penetration weld [46 and 49].

It is a partial penetration electron beam bead-on-plate (low carbon steel) weld. Traditionally the Rosenthal *2D* model would be used to analyze this weld. However, there is some heat flow through the thickness dimension since the penetration is partial and, of course, the idealized line heat source is suspect. The ellipsoidal model can be easily adapted to this weld geometry by selecting appropriate characteristic ellipsoidal parameters. A cooling time (*800* °*C* to *500* °*C*) of *1.9* seconds was measured by Chong [49] and the *FZ* and *HAZ* dimensions were reported [46].

In previous works the thermal properties and boundary conditions were usually set equal to a constant value. Convection and radiation are mostly ignored. The point, line and plane sources [37, 38 and 39] idealize a heat source which in reality is distributed. These solutions are most accurate far from the heat source. At the source, the error in temperature is large, usually infinite. Near the heat source the accuracy can be improved by matching the theoretical solution to experimental data. This is usually done by choosing a fictitious thermal conductivity value.

With numerical methods, these deficiencies have been corrected and more realistic models that are just as rigorous mathematically have been developed. Perhaps the most important factor is to distribute the heat rather than assume point or line sources. Temperature dependent thermal conductivity and heat capacity can be taken into account, Figures 2-21 and 2-22. In addition, temperature dependent convection and radiation coefficients can be applied to the boundaries. For the radiation and convection boundary conditions, a combined heat transfer coefficient was calculated from the relationship:

$$H = 24.1 \times 10^{-4} \varepsilon T^{1.61} \qquad (2\text{-}18)$$

where ε is the emissivity or degree of blackness of the surface of the body. A value of *0.9* was assumed for ε, as recommended for hot rolled steel [37].

Contact thermal resistance between the plate and the jigging can be incorporated.

Bisra [50] and Mills [52] also present thermo-physical properties for selected commercial alloys. The thermal conductivity of steels at room temperature is reduced by increasing the amount of alloy substances, Figure 2-21. This situation is limited to the phase change temperature *A1*, from where the variation disappears in the slight rise of the curve.

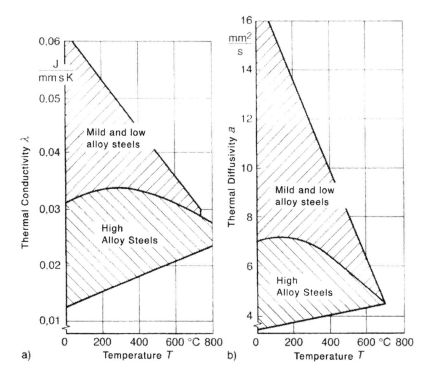

Figure 2-21: Thermal conductivity (a) and Thermal diffusivity (b) of steels as function of temperature; from [71 and 3].

The specific heat capacity for some steels as a function of temperature is shown in Figure 2-22.

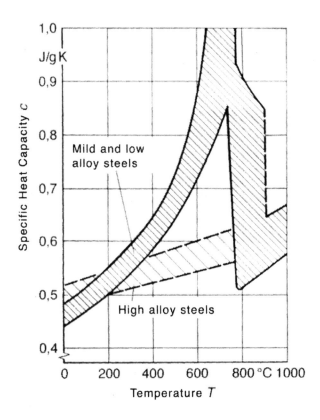

Figure 2-22: Specific heat capacity for some steels as function of temperature, latent heat at phase change temperature for ferrite pearlite (A1) and at phase change temperature for ferrit austenit (A3); from [71 and 3].

Figure 2-23 shows the density for some steels as a function of temperature.

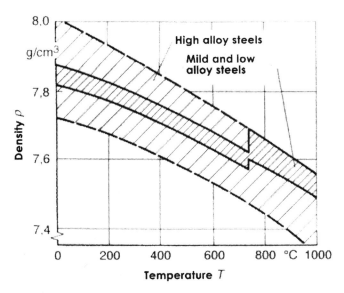

Figure 2-23: Density of some steels as function of temperature; from [71 and 3].

As shown in Figure 2-9 there are four characteristic length parameters that must be determined. Physically these parameters are the radial dimensions of the molten zone in front, behind, to the side and underneath the arc. If the cross-section of the molten zone is known from experiment, these data may be used to fix the heat source dimensions. For example, the width and depth are taken directly from a cross-section of the weld. In the absence of better data, experience suggests it is reasonable to take the distance in front of the heat source equal to one-half the weld width and the distance behind the heat source equal to twice the width. If cross-sectional dimensions are not available Christensen's expressions [48] can be used to estimate these parameters. Basically Christensen defines a non-dimensional operating parameter and non-dimensional coordinate systems. Using these expressions, the weld pool dimensions can be estimated.

The non-dimensional Christensen method was used to fix the ellipsoidal flux distribution parameters for the thick section bead on plate weld shown in Figure 2-19. The cross-sectional dimensions were reported by Chong, and the half-width dimension was applied to the flux distance in front of the electron beam heat source while the twice-width distance was applied behind the *EB*. The heat input fractions used in the computations were based on a parametric study of the model. Values of $f_f = 0.6$ and $f_r = 1.4$ were found to provide the best correspondence between the measured and calculated thermal history results.

The temperature distribution along the width perpendicular to the weld center line at 11.5 seconds after the arc passed is shown in Figure 2-24 [46]. It is compared to the experimental data from Christensen et al. [48] and the finite element analysis of the same problem by Krutz and Segerlind [42] where a disc-shaped heat source (Eq. 2-4) was used. As expected, the ellipsoidal model gives better agreement with experiment than the disc.

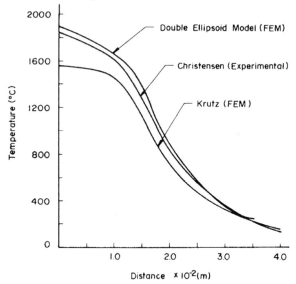

Figure 2-24: Temperature distribution along the top of the workpiece perpendicular to the weld. Experimental results of Christensen [48] compared to the computed values of Krutz and Segerlind 'disc model' [42] and the computed values using the 'double ellipsoid model' [46].

The fusion and heat affected zone boundary positions predicted by these *FEM* calculations, [46], are in good agreement with the experimental data, as shown in Figure 2-16. In addition, the *FEM* cooling times (*800 °C* to *500 °C*) are much closer to the experimental value than the cooling time calculated by the Rosenthal's analysis. The *FEM* cooling time is slightly larger than the experimental value. This may be due to neglecting the longitudinal heat flow. The radiation-convection applied to the top surface had little effect on the thermal cycle or the *FZ-HAZ* boundaries. This is to be expected for thick section welds where the heat flow is dominated by conduction.

A plot of the heat input at the surface using the double ellipsoid heat source is given in Figure 2-25, from Lindgren [22].

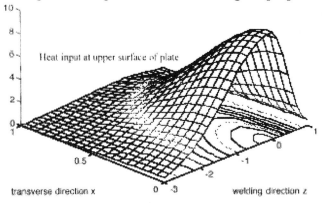

Figure 2-25: Heat input distribution using a double-ellipsoid heat source model, [22].

2.2.5 Modeling Thermal Stresses and Distortions in Welds

The thermal stress analysis of welds is more complex than the heat flow analysis because of the geometry changes and because of the complex stress-strain relationship. In designing a welding procedure, the critical issues are defects, mechanical properties, distortion and residual stress. Modeling stresses in welds includes all distortions that can be predicted by thermal stress analysis. Models of the mechanical properties of base metal, *HAZ* and weld metal, are needed as input data for thermal analysis. To evaluate the significance of a given defect, stress analysis is required.

Most thermal stress analyses have used thermo-elasto-plastic constitutive models with rate independent plasticity. Rate independent plasticity implies the viscosity is zero and therefore the relaxation time is zero. This means the stress relaxes instantly to the yield stress. The higher the temperature and the longer the time, the more important viscous deformation becomes. A rate independent model is certainly not valid in the liquid region and is suspect near the melting point where viscous effects are expected to be important; most analyses assume a cutoff temperature. They assume that the thermal strain, Young's modulus and yield strength do not change above a cutoff temperature. Temperatures above this cutoff temperature are set to the cutoff temperature for the thermal stress analysis. For steels the cutoff temperatures which have been used range from *600-800 °C* to *1100-1200 °C*.

The analysis of stress and distortion in welds involves both rate dependent plasticity and rate independent plasticity. Plastic deformation below half the melting point temperature (measured in degree Kelvin) is usually rate independent plasticity. Stress relaxation by creep or by visco-plasticity is rate dependent plasticity and it should be considered for deformation above half the melting point, in multi-pass welds and in stress-relief of welds. Rate independent plasticity is independent of the deformation rate and therefore of time (in limited deformation rates or the ratio of deformation time to the active plastic deformation). The related characteristic behaviour in a simple one dimensional model is shown in Figure 2-26a.

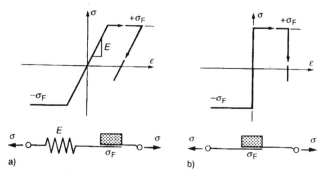

Figure2-26a, b: One dimensional model of rate independent elasto-plastic a) and ideal plastic b) of material behaviour; Young's Modulus E, with yield strength σ_F; with stress σ and strain ε, [3].

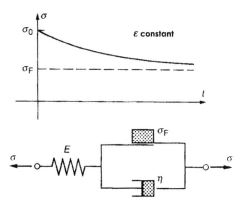

Figure 2-26c: One dimensional model of rate-dependent elasto-visco-plastic material behaviour; relaxation of beginning stress σ_0 by constant strain ε as a function of time t; Young's Modulus E, with yield strength σ_F, dynamic viscosity η, [3].

To complete this, Figure 2-26b shows an ideal plastic model which is unsuitable for the evaluation of welding stress. In contrast to this, the elasto-visco-plastic continuum defines a plastic behaviour dependent on the deformation rate and therefore on time. The related characteristic behaviour in a simple one dimensional model and the relaxation behaviour are illustrated in Figure 2-27.

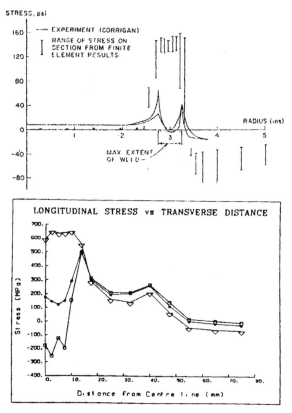

Figure 2-27: Residual longitudinal stress on the top surface vs. transverse distance from the weld centerline in HY-130 weld measured by Corrigan and computed by Hibbitt & Marcal [29] are shown in a). Note the lack of agreement between computed and measured stresses. The longitudinal stress shown in b) was computed by Oddy et al. [60] for HY-80 for three cases. The top curve ignores the phase transformation. The bottom curve includes the volume change due to the transformation. The middle curve includes the effect of transformation plasticity (In steel welds during the austenite-ferrite phase transformation the variations of stress and strain on the length scale of microstructure produce an important contribution to the plastic strain called transformation plasticity). Note the good agreement between experiment and the computed results for the middle curve.

Although the accuracy is limited by computational costs to rather coarse meshes, Figure 2-13, the residual stresses have usually agreed well with experimental results. An important exception has been the predicted residual stress in high strength steels. Hibbitt and Marcal

[29] attempted to predict the residual stress in welding *HY-130* steel. Their results were so different from experiment that they concluded that an important physical phenomenon had been missed, Figure 2-27.

In 1988, Oddy A. et al. [60] repeated the analysis but included the effect of the austenite to martensite transformation. They obtained the correct residual stress and showed that the critical phenomenon missed by Hibbitt and Marcal was transformation plasticity. This is the plastic strain that occurs during a phase change in the presence of a deviatoric stress.

Transient *3D* thermal stress analyses of welds have been described in detail in several papers [53, 56 and 65]. The thermal strain rate due to the transient temperature field induces an elastic, plastic and transformation plasticity strain rate. An exact solution to a stress analysis problem satisfies three basic laws; the conservation of linear momentum or the equilibrium equation, the constitutive relation between stress and strain and the compatibility relations between strain and displacements which is the conservation of mass. In addition it satisfies two types of boundary conditions; the prescribed displacements or essential boundary conditions and the prescribed tractions or natural boundary conditions. A displacement *FEM* formulation is used to solve the constitutive, compatibility and equilibrium equations. The displacements, rotations and strains are large, transient and *3D*. The thermal strain must be modeled with care (Oddy et al [65]) to avoid an incompatibility between thermal strain and strain from the displacement field.

Residual stress and distortion of welds are strong functions of plasticity. We are more concerned with the neighborhood of arc welds that have a small pool of liquid metal as distinct from the structure being welded.

Near the weld pool the kinematics is definitely *3D*. Farther from the weld pool, the kinematics in appropriate structures sometimes can be modeled with plane or shell models. The plane strain model was often used in the past, because it is computationally cheap.

During a phase transformation, the volume change induces a local variation in the stress field. For example, during the austenite to martensite transformation, the volume change could place a martensite grain in compression and some of the surrounding austenite in tension just at its yield stress. If an average tensile stress is superimposed upon this microscopic stress field, the austenite could now yield in tension with an infinitesimal applied tensile stress. Now reverse the order. Apply an infinitely small macroscopic tensile stress first. Then transform some austenite to martensite to reach the yield stress of austenite locally. An additional transformation will cause plastic strain. This additional plastic strain is defined to be transformation plastic strain. It will stop when the transformation stops. It arises from the interaction between the microscopic and macroscopic stress fields. This two-scale or multi-scale is a bit similar to turbulence models in fluid flow that also use an average or macroscopic velocity and velocity fluctuation. If the macroscopic stress field is not modeled (i.e., the microscopic stress field is modeled directly), then classical plastic strain describes the deformation, and transformation plasticity does not occur. While the classical plastic strain rate is proportional to the deviatoric stress rate, the transformation plasticity strain rate is proportional to the product of the deviatoric stress times the transformation rate of the phase change. It is not creep because it is not proportional to time. Leblond's exposition of this theory is beautiful [57].

The high temperature range ($T \geq 0.5T_m$) is usually considered to be of minor importance in generating residual stress due to the low value of yield stress. However it is important in hot cracking that is caused by the high temperature strain close to the molten temperature. The geometric changes at high temperatures are inherited by regions that cool from high temperatures. These geometric changes can be very important. The thermo-mechanics of causing the residual stress by welding is shown schematically in Figure 2-28. In this figure the scheme of plastic zone distribution for a case of quasi-stationary temperature field caused by a moving line heat source is presented. The parabola-like curve drawn as a broken line separates the region being heated in front of the curve and the region being cooled behind the curve. The region being heated tends to be in compression and the region being cooled tends to be in tension. The elastic unloading zone shown as a strip separates the compressed and tensioned areas. For different points in the area the schematics highlight the strain-stress cycle related to the local load cycle, without considering temperature dependency.

As an example, adopted from Pilipenko [18], the point 6 in Figure 2-28 may be suggested. At first sight the schematic stress-strain cycle in point 6 should have looked like point 5. But the difference is that material in point 6, after reaching some elastic and plastic compression, was 'annealed' inside of the material softening isotherm. Points 1, 2 and 3 represent the evolution of stress developed at a point lying at some distance from the weld centreline. First the material is being exposed to elastic compression (point 1), and then, reaching the yield limit, the material undergoes plastic deformation (point 2), followed by elastic unloading (point 3). Point 7 has a peculiar position. It lies on the weld centreline and the material in this point has been subjected only to elastic and then plastic tensioning.

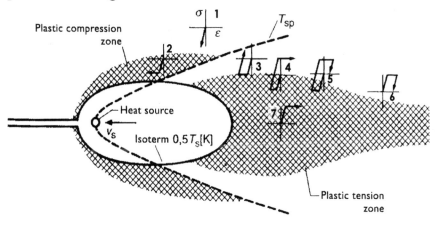

Figure 2-28: Plastic compression and tension zones; local stress-strain cycles in quasi-stationary temperature field of the moving heat source, from Radaj [3].

The temperature of the metal in welds varies from the boiling temperature to room temperature. The domain includes the liquid weld pool and the far field of solid near room temperature. The liquid is well described as a Newtonian fluid characterized by a temperature dependent viscosity. The temperature of the solid can be considered to be a linear viscous material characterized by a viscosity due to the diffusion of dislocations. From temperatures above $T_{VP} \approx 0.5 or 0.8 T_m$, the solid can be considered to be visco-plastic and characterized by an elasticity tensor, viscosity and deformation resistance. Below T_{VP}, the solid can be considered to be a rate independent plastic material characterized by an elasticity tensor, yield strength and isotropic hardening modulus. Thus the constitutive equation of a material point must be able to change type with space and time during the welding process. The constitutive model is assigned to be rate independent if the temperature is less than T_{VP}, rate dependent if the temperature is in a range *0.5-0.8 T_m* and linear viscous if the temperature is greater than *0.8T_m*, where T_m is a melting point or solidus temperature in degrees of Kelvin. The constitutive model type changes as a function of temperature in space and time.

The practical application of a weld simulation with different constitutive equations within the above mentioned temperature areas will be discussed in detail in chapter *V*.

An example of computed displacements in a weld is shown in Figure 2-29.

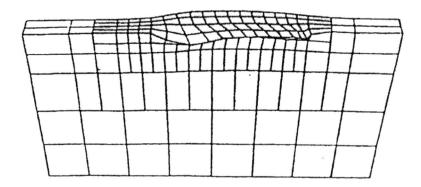

Figure 2-29: Displacements computed by a thermo-elasto-plastic analysis for a 1.6 kJ/mm edge weld on a carbon steel bar 12.7mm in width. Note that the displacements have been magnified by a factor of 50 to make them discernable. Note the displacements in the longitudinal direction that violate the assumptions of plane strain.

2.2.6 Microstructure Modeling in Heat Affected Zone (HAZ)

Many failures of welds initiate in the heat affected zone adjacent to the weld metal, e.g., in the coarse grained *HAZ*. For this reason, welding engineers devote much of their effort to controlling the microstructure and hence toughness of the *HAZ*. Metallurgists usually describe the evolution of microstructure in low alloy steels with isothermal transformation diagrams or continuous cooling transformation *(CCT)* diagrams. These diagrams are maps that depict the starting time and temperature of the various microstructural transformations that occur over a range of cooling trajectories. These diagrams have been developed for heat treating. Heat treating usually assumes an equilibrium microstructure at a soaking temperature of the order of *900 °C*. The Jominy test, that imposes a jet of cold water on one end of a bar, is a physical model of the evolution of microstructure in heat treating.

This *CCT* model cannot be applied directly to welding for several reasons. The peak temperature in the *HAZ* of welds ranges from a bit below the eutectoid temperature to the melting point. Grain growth in the austenite phase is non-uniform in the *HAZ*. It is sensitive to the time and temperature curve and the presence of precipitates such as *NbC* that inhibit grain growth. Also the *CCT* diagrams are not a convenient representation for a computational model.

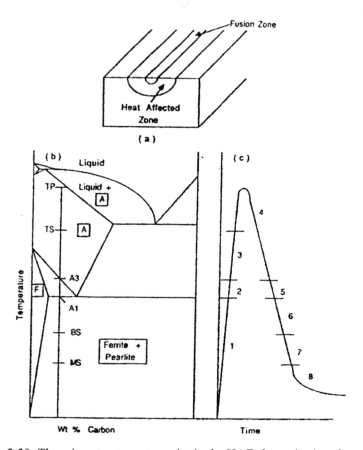

Figure 2-30: The microstructure at a point in the HAZ shown in a) under goes the thermal cycle shown in c). The microstructure evolves through the following stages: 1) ferrite and pearlite are at equilibrium; 2) ferrite and pearlite transform to austenite; 3) austenite grain growth is inhibited by NbC and VC; 4) austenite grain growth occurs; 5) austenite decomposes to ferrite; 6) austenite decomposes to pearlite; 7) austenite decomposes to bainite; 8) austenite decomposes to martensite.

For these reasons, the model developed by Kirkaldy et al.[61] for Jominy bars, was extended by Watt et al.[66] and Henwood et al.[62] to model the *HAZ* of welds in low alloy steels. Input data to this model are the composition of base metal, initial microstructure and the transient temperature computed from *FEM*. The output is the fraction of ferrite, pearlite, austenite, bainite, martensite and the grain size of the austenite at each point *(x,y,z,t)*.

The model begins by computing the *Fe-C* equilibrium diagram for this composition, Figure 2-30b.

For each time step in the transient analysis, the phase fractions at the start are initial data and the phase fractions at the end of the step are computed. On heating, phase transformations are assumed to occur so quickly that the kinetic effects can be ignored, i.e., local equilibrium is assumed. In effect, superheating is ignored. After ferrite and pearlite have fully transformed to austenite, austenite grain growth begins after the dissolution of *NbC* and/or *VC* if they are present. Austenite grain growth is evaluated by integrating an ordinary differential equation *(ODE)*. Grain growth continues until the *A3* temperature is reached on cooling. At this time the decomposition of austenite to a) ferrite, b) pearlite and c) bainite is modeled with a sequence of *ODEs*. Finally, if austenite is still present at the martensite start temperature, austenite is transformed to martensite by an algebraic equation proposed by Koisten and Marburger [63].

The *ODE* for austenite to ferrite is shown in equation (2-19.):

$$\frac{dX}{dt} = \frac{2^{\frac{G-1}{2}}(\Delta T)^3 \exp\left(-\dfrac{23500}{RT}\right)}{59.6Mn + 1.45Ni + 67.7Cr + 24.4Mo} X^{\frac{2(1-X)}{3}}(1-X)^{\frac{2X}{3}} \quad (2\text{-}19)$$

The exponent containing *G*, the austenite grain size, in the first term reflects the fact that since ferrite nucleates at the austenite grain boundary, the density of nuclei is a function of the austenite grain boundary area. The next term reflects the metal physics experiments that shows the rate of growth of ferrite increases as *ΔT^3* where *$\Delta T=A3-T$* is the undercooling temperature for the austenite-ferrite transformation. This is due to the increasing difference in free energy between austenite and ferrite with under-cooling. The exponential term reflects the decrease in the diffusivity of carbon in iron with temperature. The denominator reflects the effect of alloying elements on diffusivity. The last term is the predator prey factor which states that the rate of production of ferrite is a product of the fraction of ferrite present, X, and the fraction of austenite present (1-X).

$$X = \frac{X_F}{X_{FE}} \qquad (2\text{-}20)$$

where X_F is the fraction of ferrite formed and X_{FE} is the equilibrium amount of ferrite.

$$G = 1 + 1.44[\ln((\frac{25.4}{g})^2 \times 100)] \qquad (2\text{-}21)$$

where g is the grain size in microns.

Kirkaldy has proposed the following *ODE* relation for the austenite decomposition to pearlite and it is similar to equation (4-21):

$$\frac{dX}{dt} = \frac{2^{\frac{(G-1)}{2}} (\Delta T)^3 D}{1.79 + 5.42(Cr + Mo + 4MoNi)} X^{\frac{2(1-X)}{3}} (1-X)^{\frac{2X}{3}} \qquad (2\text{-}22)$$

where ΔT is the undercooling given as $(A_1 - T)$ and D is a diffusion parameter and it is given by the following relation, adopted from [74]:

$$\frac{1}{D} = \frac{1}{\exp(-\frac{27500}{RT})} + \frac{0.01Cr + 0.52Mo}{\exp(-\frac{3700}{RT})} \qquad (2\text{-}23)$$

and

$$X = \frac{X_P}{X_{PE}} \qquad (2\text{-}24)$$

where X_P is the fraction of pearlite formed and X_{PE} is the equilibrium amount of pearlite. The equilibrium value of ferrite is calculated at each time step. Therefore the equilibrium amount of ferrite at any specific temperature is the maximum amount of ferrite which can be formed from austenite. The remaining amount of austenite is available for the formation of pearlite. This allows the value of X_{PE} to be set equal to $1 - X_{FE}$.

The *ODE* governing decomposition of austenite to bainite is taken from the work of Kirkaldy, adopted from Khoral [74]:

$$\frac{dX}{dt} = \frac{2^{\frac{(G-1)}{2}} (\Delta T)^2 \exp(-\frac{27500}{RT})}{10^{-4}(2.34 + 10.1C + 3.8Cr + 19Mo)Z} X^{\frac{2(1-X)}{3}} (1-X)^{\frac{2X}{3}} \qquad (2\text{-}25)$$

where X is the amount of bainite formed and ΔT is the undercooling given as *(BS-T)*. Z is given by the following relation:

$$Z = \exp[X^2(1.9C + 2.5Mn + 9Ni + 1.7Cr + 4Mo - 2.6)] \qquad (2\text{-}26)$$

The value of Z is set to 1.0 if: (1.9C+2.5Mn+9Ni+1.7Cr+4Mo-2.6)<0.

The martensitic transformation is given by the following relation taken from the work of Koistenen and Marburger [63]:

$$X_M = 1 - \exp[-k_1(MS - T)] \qquad (2\text{-}27)$$

where X_M is the volume fraction of martensite formed, MS is the martensite start temperature, T is the instantaneous temperature and k_1 is a constant and its value for most steel types is $0.011°C^{-1}$.

This microstructure model does not apply to the weld metal because the effects of solidification are ignored. Bhadeshia [64] has developed models that could be applied to weld metal.

2.2.7 Spatial Integration Schemes

Since the Finite Element Method *(FEM)* is primarily an exercise in numerical integration, it is not surprising that better integration schemes are being sought. Numerical integration schemes replace an integral by a summation; e.g. $\int_{-1}^{1} f(x)dx = \sum_{i=1}^{n} w_i f(x_i)$. The choice of the number of sampling points, i, the weights, w_i, and the location of the sampling points, x_i, characterize the integration scheme. If $f(x)$ is a polynomial of degree $2n\text{-}1$, it can be integrated exactly with n sampling or integration points as shown by Gauss. In two dimensions, the stiffness matrix of the popular *4* node rectangular element (but not a quadrilateral) can be integrated exactly with four integration points. If only one integration point is used, the integration is approximate and spurious nodes may appear that corrupt the solution. Belytschko [16] discovered a way to stabilize these spurious nodes and thus reduce the integration costs by almost a factor of four for 4-node quads and almost 8 for 8-node bricks.

References

1. Finger S. and Dixon J.R. A review of research in mechanical engineering design. Part I: descriptive, prescriptive and computer-based models of design process. Research in Engineering Design, 1989 1, pp 51-57
2. Finger S. and Dixon J.R. A review of research in mechanical engineering design. Part II: representation, analysis and design for the life cycle, Research in Engineering Design, No.1, p 121-137, 1989
3. Radaj D. Eigenspannungen und Verzug beim Schweissen, Rechen- und Messverfahren, Fachbuchreihe Schweisstechnik, DVS-Verlag GmbH, Duesseldorf 2000
4. Runnemalm H (1999). Efficient finite element modeling and simulation of welding. Doctoral Thesis, Lulea.
5. Chihoski Russel A. Understanding weld cracking in Aluminum sheet, Welding Journal, Vol. 25, pp 24-30, Jan. 1972
6. Requicha A.A.G.and Voelcker H.B. Solids modeling; current status and research directions, IEEE Computer Graphics and Application, Vol. 7, pp 25-37, 1983
7. Kela A., Voelcker H. and Goldak J.A. Automatic generation of finite element meshes from CSG representations of solids, International conference on accuracy estimates and adaptive refinements in finite element computations (ARFEC), sponsored by the International Association of Computational Mechanics, Lisbon Portugal June 19-20,1984
8. Sheperd M.S. and Law K.H. The modified-quadtree mesh generator and adaptive analysis: International conference on accuracy estimates and refinements in finite element computations (ARFEC), sponsored by the International Association of Computational Mechanics, Lisbon Portugal, June 19-20, 1984
9. Donnea J. Arbitrary Lagrangian-Eulerian finite element methods, Computational Methods for Transient Analysis, vol. 1, North-Holland publishing Co, Edited by Belyschko T. and Hughes T.J.R., p 473-516, 1983
10. McDill JMJ, Goldak JA, Oddy AS, and Bibby MJ. Isoparametric quadrilaterals and hexahedrons for mesh-grading elements, Comm. Applied Nu. Methods, Vol. 3, pp 155-163, 1987
11. Swanson J.A., Cameron G.R. and Haberland J.C. Adapting the Ansys finite element program to an attached processor, IEEE Computer Journal, 1983, Vol. 16, No 6, pp 85-91
12. Glowinski R., Mantel B., Periaux J. Perrier P. and Pironneau: On an efficient new preconditioned conjugate gradient method. Application to the in-core solution of the Navier-Stokes equations via non-linear least-square and finite element methods , Finite Elements in Fluids, Vol. 4, pp 365-401, 1982
13. Ajiz M.A. and Jennings A. A robust incomplete choleski-cojugate gradient algorithm, International Journal for Numerical Methods in Engineering, Vol. 20, pp 949-966, 1984

14. Hughes T.J.R., Levit I. and Winget J. An element-by-element solution algorithm for problems of structural and solid mechanics, Computer Methods in Applied Mechanics and Engineering, Vol. 36, pp 241-254,1983
15. Hughes T.J.R., Winget J., Levit I. and Tezduyar T.E. New alternative direction procedures in finite element analysis based upon EBE approximate factorizations, ASME Proc. of Symp. On Recent Developments in Nonlinear Stress and Solid Mechanics, June 1983
16. Belyschko T. An overview of semidiscretization and time integration procedures, Computational Methods for Transient Solid Mechanics, June 1983
17. Lui W.K. Mixed time integration methods for transient thermal analysis of structures, Nasa Contractor Report 172209, pp 1-55, Sept. 1983
18. Pilipenko A (2001) Computer simulation of residual stress and distortion of thick plates in multielectrode submerged arc welding. Doctoral thesis, Norwegian University of Science and Technology.
19. Mahin K.W.,MacEwen S., Winters W. S., Mason W., Kanouff M. and Fuchs E.A. Evaluation of residual stress distortions in a traveling GTA weld using finite element and experimental techniques, Proc. of Modeling of Casting and Welding Processes IV, Ed. Giamei A.F. and Abbaschian G.R., pp 339-350, April 17-22, 1988, Pub. Eng. Foundation and TMS/AIME
20. Goldak J. A., Bibby M.J., Downey D. and Gu M. Heat and fluid flow in welds, Advanced Joining Technologies, Proc. of the International Institute of Welding Congress on Joining Research, July 1990
21. Tall L. Residual stresses in welded plates-a theoretical study, Welding Journal, Vol. 43, No. 1, pp 10s-23s, 1964
22. Lindgren L-E. Finite element modeling and simulation of welding Part I Increased complexity, J of Thermal Stresses 24, pp 141-192, 2001
23. Tsuji I. Transient and residual stresses due to butt-welding of mild steel plates, Memoirs of the Faculty of Engineering, Kyushu University, vol. XXVII, No. 3. 1967
24. Iwaki T. and Masubuchi K. Thermo-elastic analysis of orthotropic plastic by the finite element method, J Soc. Naval Arch. Japan, Vol. 130, pp 195-204, 1971
25. Ueda Y. and Yamakawa T. Thermal stress analysis of metals with temperature dependent mechanical properties, Proc. int. Conf. Mech. Behavior of Materials, p 10, 1971
26. Ueda Y. and Yamakawa T. Analysis of thermal elastic-plastic stress and strain during welding by finite element method, JWRI, Vol. 2, No. 2, pp 90-100, 1971
27. Ueda Y. and Yamakawa T. Mechanical cracking characteristics of cracking of welded joints, Proc. of 1st Int. Symposium on Precaution of Cracking in Welded Structures Based on Resent Theoretical and practical Knowledge, The Japan Welding Society, Tokyo Japan, p IC5.1, 1971

28. Fujita Y., Takesshi Y., Kitamura M. and Nomoto T. Welding stresses with special reference to cracking, IIW Doc X-655-72, 1972

29. Hibbitt H.D. and Marcal P.V. A numerical thermo-mechanical model for the welding and subsequent loading of a fabricated structure, Computers & Structures, Vol. 3, pp 1145-1174, 1973

30. Friedman E. Thermomechanical analysis of the welding process using the finite element method, ASME J. Pressure Vessel Technology, Vol. 97, No. 3, pp 206-213, Aug. 1975

31. Fujita Y., Takeshi Y. and Nomoto T. Studies on restraint intensity of welding, IIW Doc X-573-70,1970

32. Fujita Y. and Nomoto T. : Studies on thermal stresses with special reference to weld cracking, 1st Int. Symposium on Precaution of Cracking in Welded Structures Based on recent Theoretical and Practical Knowledge, The Japan Welding Society, p IC6.1, 1971

33. Mahin K., Shapiro A.B., Hallquist J. Assessment of boundary conditions limitations on the development of a general computer Model for fusion welding , Proc. of an International Conf. on Trends in Welding Technology, pp 215-224, Gatlinburg Tennessee USA, May 18-22, 1986, ed. David S. A.

34. Tekriwal P., Mazumder J. Finite element modeling of the arc welding process, Proc. of an international Conf. on Trends in Welding Technology, pp 71-80, Gatlinburg Tennessee USA, May 18-22, 1986, ed. David S. A.

35. Goldak JA, McDill JMJ, Oddy AS, House R, Chi X and Bibby MJ. Computational heat transfer for Weld Mechanics, Proc. of an International Conf. on Trends in Welding Technology, pp 15-20, Gatlinburg Tennessee USA, May 18-22, 1986, ed. David S. A.

36. Goldak J. A., Patel B., Bibby M.J. and Moore J.E. Computational weld mechanics, Invited opening paper for AGARD Workshop-Structures and Materials 61st Panel Meeting, Oberammergau, Germany, pp 1-1 1-32, Sep. 8-13, 1985

37. Rykalin R.R. Energy sources for welding, Welding in the World, Vol. 12, No. 9/10, p 227-248, 1974

38. Rosenthal D.: the theory of moving sources of heat and its application to metal treatments, Trans ASME, Vol. 68, p 849-865, 1946

39. Myers P.O., Uyehara O.A. and Borman G.L. Fundamentals of heat flow in welding, Welding Research Council Bulletin, New York No. 123, 1967

40. Pavelic V., Tanbakuchi R., Uyehara O. A. and Myers: Experimental and computed temperature histories in gas tungsten arc welding of thin plates, Welding Journal Research Supplement, Vol. 48. pp 295s-305s, 1969

41. Andersson B.A.B. Thermal stresses in submerged-arc welded joint considering phase transformation, Journal of Engineering Materials and Technology, Trans. ASME, Vol. 100, pp 356-362, 1978

42. Krutz G.W. and Segerlind L.J. Finite element analysis of welded structures, Welding Journal Research Supplement, Vol. 57, pp 211s- 216s, 1978

43. Friedman E. Thermo-mechanical analysis of the welding process using the finite element method, Journal Pressure Vessel Technology, Trans. ASME, Vol. 97, No 3, pp 206-213, 1975

44. Paley Z. and Hibbert P.D. Computation of temperatures in actual weld designs, Welding Journal Research Supplement, Vol. 54, pp 385s-392s, 1975

45. Westby O. Temperature distribution in the workpiece by welding, Department of Metallurgy and Metals Working, The Technical University, Trondheim Norway, 1968

46. Goldak J., Chakravarti A. and Bibby M. A finite element model for welding heat sources, Metallurgical Transactions B, Vol. 15B, pp 299-305, June 1984

47. McDill JM, Oddy AS. and Goldak JA. An adaptive mesh-management algorithm for three-dimensional automatic finite element analysis, Transactions of CSME, Vol. 15, No 1, pp 57-69, 1991

48. Christensen N., Davies L.de.V. and Gjermundsen: British Welding Journal, Vol. 12, pp 54-75, 1965

49. Chong L.M. (1982). Predicting weld hardness. Master's Thesis, Carleton University.

50. The British Iron and Steel Research Association, Physical Constants of Some Commercial Steels at Elevated Temperatures, London Butterworths Scientific Publications, 1953

51. Ohji T. Physics of welding (1)- Heat conduction theory and its application to welding. Welding Intern. 8 (1994) 12, 938-942 trans. From J Japan Welding Soc. 63 (1994) 4, 32-36

52. Mills Kenneth C. Recommended values of thermo physical properties for selected commercial alloys, Woodhead Publishing Ltd ISBN 1 85573 569 5, ASM International ISBN 0-87170-753-5, 2002

53. Goldak J.A., Breiguine V., Dai N., Hughes E. and Zhou J. Thermal Stress Analysis in Solids Near the Liquid Region in Welds. Mathematical Modeling of Weld Phenomena, 3 Ed. By Cerjak H., The Institute of Materials, pp 543-570, 1997

54. Goldak J.A. et al. Coupling heat transfer, microstructure evolution and thermal stress analysis in weld mechanics. IUTAM Symposium, Lulea Sweden, 1991

55. Weber G. and Anand L. Finite deformation constitutive equations and a time integration procedure for isotropic, hyperelastic-viscoplastic solids, Comput. Methods Appl. Mech. Eng., Vol. 79, pp 173-202, 1990

56. Oddy AS, Goldak JA and McDill JMJ. Numerical analysis of transformation plasticity relation in 3D finite element analysis of welds, European Journal of Mechanics, A/Solids, Vol. 9, No. 3 pp 253-263, 1990

57. Leblond J.B., Mottet G. and Devaux J.C. A theoretical and numerical approach to the plastic behavior of steels during phase transformation-II, Study of classical plasticity for ideal-plastic phases, J. Mech. Phys. Solids, Vol. 34, pp 411-432, 1986

58. Gu M. and Goldak J.A. Steady state formulation for stress and distortion of welds, J. of Eng. For Industry, Vol. 116, pp 467-474, Nov. 1994
59. Goldak J. Keynote address modeling thermal stresses and distortions in welds, Resent Trends in Welding Science and Technology TWR,89, Proc. of the 2nd International Conference on Trends in welding Research, Gatlinburg Tennessee USA,14-18 May 1989
60. Oddy A. Goldak J.A. and McDill J.M.J. Transformation effects in the 3D finite element analysis of welds, Proc. of the 2nd International Conf. on trends in Welding research, Gatlinburg Tennessee USA, May 1989
61. Kirkaldy J.S. and Venugopalan: In phase transformations in ferrous alloys, ed. By Marder A.R. and Goldenstein J.I., AM. Inst. Min. Engrs., Philadelphia, Pa, pp 125-148, 1984
62. Henwood C. Bibby M.J., Goldak J.A. and Watt D.F. Coupled transient heat transfer microstructure weld computations, Acta Metal, Vol. 36, No. 11, pp 3037-3046, 1988
63. Koisten D.P. and Marburger R.E. A general equation prescribing the extent of the austenite-martensite transformation in pure iron-carbon alloys and plain carbon steels, Acta Met., Vol. 7, pp 59, 1959
64. Bhadeshia H. Modeling the microstructure in the fusion zone of steel weld deposits, Proceedings of an International Conference on Trends in Welding Technology, Gatlinburg Tennessee USA, ed. David S. A., May 15-19, 1989
65. Oddy A.S., McDill J.M. and Goldak J.A. Consistent strain fields in 3D finite element analysis of welds, J. Pressure Vessel Tech., Vol. 25, No 1, pp 51-53, 1990
66. Watt D. F., Coon L., Bibby M. J., Goldak J.A. and Henwood C. Modeling microstructural development in weld heat affected zones, Acta Metal, Vol. 36, no 11, pp 3029-3035, 1988
67. Gu M. (1992). Computational weld analysis for long welds. Doctoral thesis Carleton University.
68. Bibby M.J., Shing G.Y. and Goldak J.A. A model for predicting the fusion and heat affected zone sizes of deep penetration welds, CIM Metallurgical Quarterly, Vol. 24, No. 2, 1985
69. Chakravarti A.P., Goldak J.A. and Rao A.S. Thermal analysis of welds, International Conference on Numerical Methods in Thermal Problems, Swansea UK, July 1985
70. Key J.F., Smartt H.B., Chan J.W. and McIlwain M. E. Process parameter effects on the arc physics and heat flow in GTAW, p 179-199
71. Richter F. Die wichtigsten physikalischen Eigenschaften von 52 Eisenwerkstoffen. Stahleisen-Sonderberichte No. 8, Publ. Stahleisen, Duesseldorf 1973
72. Sudnik W.A. and Erofeew W.A. Rastschety swarotschnych processov na iwm Techn. Univ. Tula, Sweden, 1986

73. Ohji T. Ohkubo A. and Nishiguchi K. Mathematical modeling of molten pool in arc welding, Mechanical Effects of Welding, p 207-214, Publ. Springer, Berlin 1992

74. Khoral P (1958) Coupling microstructure to heat transfer computation in weld analysis. Masters Thesis, Carleton University.

75. Okerblom, N.O. The calculations of deformations of welded metal structures, London, Her Majesty's Stationary Office, 1958

76. Vinokurov, V. A. Welding stresses and distortion, The British Library Board, 1977

Chapter III

Thermal Analysis of Welds

3.1 Introduction and Synopsis

The conservation of energy is the fundamental principle in thermal analysis. Therefore in heat transfer theory, we are concerned with energy and ignore stress, strain and displacement. The principal phenomena are shown in Figure 3-1.

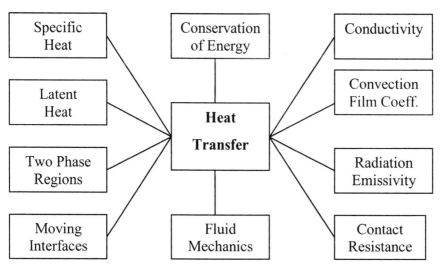

Figure 3-1: The important phenomena that should be captured in models for heat transfer analysis are shown above

The temperature field in the neighborhood of weld pool can be obtained by specifying the heat input. We will see that all weld heat source models based solely on the heat equation, specify either the source term Q, or choose a boundary of the domain near the weld pool and prescribe the temperature or prescribe the flux q on that boundary. Each of these must be specified as functions of time and position. While one can choose various types of functions, e.g., Gaussian, polynomials, Fourier series, Legendre polynomials, spherical harmonics, etc., to represent the function, which type of function is chosen is in a sense secondary. It is secondary in the sense that any function can be expressed as a linear combination of functions in any complete function space. For example, any reasonable function can be expressed in terms of a Fourier series. In a sense this is similar to the point that any problem that can be posed and solved in a Cartesian coordinate system, can also be solved in other coordinates systems such as cylindrical, spherical, ellipsoidal, toroidal, etc. Some choices may be more convenient and the equations may look simpler and could be easier to solve. However, the fact remains that if the problem can be solved in one coordinate system, it can be solved in any other equivalent coordinate system.

When we discuss a model for a heat source, it is worth distinguishing what is intrinsic to the model, i.e., what is true for all coordinate systems, and what depends on the representation and on the choice of a particular coordinate system. This is not to argue that the choice of coordinate system, e.g., Cartesian or cylindrical, and the choice of function space, e.g., Fourier series or Gaussian functions, does not matter. It is to argue that the time to make this choice is after choosing the intrinsic model.

If there are no phase changes and the thermal conductivity and specific heat are strictly positive, then given initial conditions, boundary conditions and a domain with a reasonably smooth boundary, one can show mathematically that the solution of the heat equation exists and is unique. If there are phase changes, such as dendrite formation during solidification, then the solution exists but need not be unique. One reason for this lack of uniqueness is that we cannot know, at least not yet, the distribution of nuclei. Another is that dendrite branching might not be unique.

Ignoring for the moment the question of how accurately the problem can be solved, let us note that there are only a small number of parameters involved in the heat equation. Therefore the number of types of intrinsic heat source models that could be developed is very small.

In any real weld, one can argue from physics that a real temperature field does exist. If one had sufficiently accurate sensors, this real temperature field could be measured accurately. If this temperature field is sufficiently smooth, say continuous with a piecewise continuous spatial temperature gradient, then the energy equation can be written in a variational form. This is the form usually used in finite element analysis *(FEM)*. If second derivatives are piecewise continuous, then the energy equation can be written in the form of a partial differential equation *(PDE)*. Finite difference methods *(FDM)* are usually based on the *PDE* form. Normally, a thermal problem is posed by specifying the domain, material parameters such as thermal conductivity, specific heat, latent heats, initial conditions and boundary conditions. Then the solution of this problem is the temperature and phase fractions at all points (x,y,z,t) in the space-time domain. Here we have used the notation of a Cartesian coordinate system for space and time but any coordinate system will serve equally well. If we ignore phenomena such as dendrite formation, then it is well known that this solution is unique.

If the reader has accepted our assumptions that in a region of space that we call a domain, the solution exists and is unique, then the problem can be posed in several equivalent forms. Note that here our assumption of existence and uniqueness is based on physics not mathematics. In particular, if we know the solution, i.e., temperature as a function of (x,y,z,t), then we can evaluate the energy equation directly to determine, initial conditions for any time, boundary conditions for any choice of a sub-domain of the full domain, and distributed heat input for any region. The most obvious choices are that the boundary conditions can be stated either in terms of fluxes, i.e., Neumann Boundary Conditions *(BCs)*, or in terms of prescribed temperatures, i.e., Dirichlet *BCs*. If the distributed heat input is not zero and is in the domain, then the value given by the heat equation must be used as is. There is no alternative. A very important point is that the thermal problem can be solved on any sub-domain of the full domain. We emphasize that for all of these choices, only one solution exists. Here there is no notion of approximation or error analysis involved. Here we consider only exact solutions. At this point, the various choices appear to have equal value.

3.2 Heat Transfer Theory

For the moment we will assume that the velocity of the material points is zero. However, the arc is allowed to move with some velocity. Also an observer is allowed to move with some velocity. We will state the mathematics more precisely later but here we simply want to present an intuitive view of terms or mechanisms in the heat equation or conservation of energy. Within this fundamental principle there is a concept of a flux of energy with units of (W/m^2). The flux q is proportional to thermal conductivity κ ; a material property, times the temperature gradient ∇T, i.e., $q = -\kappa \nabla T$. For $\kappa > 0$, this enforces the 2nd law of thermodynamics and the Clausien-Duhem inequality. There is the fundamental notion of energy per unit volume called specific enthalpy h (sometimes it is more convenient to use specific enthalpy per unit mass.). If there are no phase changes, then for a small change of temperature, the increment in specific enthalpy dh at constant pressure can often be expressed as $dh = c_p \, dT$ where c_p is a material property and dT is an increment in temperature.

More generally, the specific enthalpy at a given temperature and a set of i phases, with volume fraction f_i for phase i is given by the integral:

$$h\left(T_2, \sum_{i=1}^{Phases}\right) = (h_{T_1, f_{i1}} + \int_{T_1, f_{i1}}^{T_2, f_{i2}} cp\,dT + \int_{T_1, f_{i1}}^{T_2, f_{i2}} L_i df_i) \qquad (3\text{-}1)$$

However, in words, for a given temperature and a set of phase fractions the increment in specific enthalpy is the sum of increments due to temperature changes and increments due to latent heat generated by phase changes. The enthalpy is measured with respect to some chosen reference state. There can also be heat generated per unit volume Q with units of (W/m^3). Most of the work of plastic deformation is converted to such heat source Q.

The heat equation expressed as a partial differential equation holds in the neighborhood of every point in a region of Euclidean space that mathematicians call a domain. On the boundaries of the domain, one must specify the interaction of the domain with the universe external to the domain. This is done by specifying boundary conditions. Basically, there are two choices for boundary conditions. One choice is to specify the flux on the boundary. This is the natural choice to describe the action of a burner or arc on a plate. Also if a section of the boundary is insulated so that the flux is zero, then specifying zero flux is natural. The other choice is to specify the temperature on the boundary. This is a natural choice for a plate partially immersed in an `ice bath' with constant temperature. At each point on the boundary one must choose to specify either the flux or the temperature. For transient problems in which the temperature and phase fractions are functions of both space (x,y,z) and time t, one must state the conditions at time zero, t_o for every point in the domain. These are called initial conditions.

Now we can state the heat equation for a material point in the domain:

$$\frac{Dh}{Dt} + \nabla \cdot \left(- \kappa \nabla T\right) + Q = 0 \qquad (3-2)$$

The first term is the rate of change of specific enthalpy. The second term is the flow of heat out of or into the neighborhood of this point due to the flux. Mathematically it is the divergence of the flux $\nabla \cdot q = \nabla \cdot \left(- \kappa \nabla T\right)$ (W/m^2). The last term is Q the source of heat generation term with units of (W/m^3).

One sees that there are only these three terms in the heat equation (with zero velocity) that can be `manipulated' in addition to the associated boundary conditions of either prescribed flux or prescribed temperature type. Therefore any model of a weld heat source based on the heat equation must specify something about one or more terms in this heat equation or its *BCs*.

Suppose that a heat source is moving at a constant speed v in the positive x-direction. We can define an Eulerian (moving) frame with origin at the center of the source, and coordinates *(ξ ,y ,z)* .The transformation from *(x, y, z)* to *(ξ ,y ,z)* is given by: *ξ= x – v t*, where *t* is time. Noting that the material time derivative *T* in equation (3-2) becomes:

$$T = T_t + \frac{\partial \xi}{\partial t} T_\xi = T_t - vT_\xi \tag{3-3}$$

We obtain the equation of conservation of energy in the Eulerian frame:

$$T_{\xi\xi} + T_{yy} + T_{zz} + \frac{Q}{K} = -\alpha vT_\xi + \alpha T_t \tag{3-4}$$

Here, T_t represents the time derivative of temperature at a point fixed with respect to the heat source. In a steady state, this derivative is zero, and so equation (3-4) becomes:

$$T_{\xi\xi} + T_{yy} + T_{zz} + \frac{Q}{\kappa} = -\alpha vT_\xi \tag{3-5}$$

When *α* and *v* are constant, equation (3-5) can be solved more easily by applying the transformation:

$$T = T_0 + e^{-(\alpha v \xi)/2} \varphi(\xi, y, z) \tag{3-6}$$

Substituting this into equation (3-5), we have:

$$\varphi_{\xi\xi} + \varphi_{yy} + \varphi_{zz} + \frac{e^{(\alpha v)/2} Q}{\kappa} = \frac{(\alpha v)^2}{4} \varphi \tag{3-7}$$

$$\nabla^2 \varphi - \frac{(\alpha v)^2}{4} \varphi = -e^{(\alpha v)/2} \frac{Q}{\kappa} = \hat{Q}$$

Equation (3-7) is symmetric positive definite. After properly transforming the boundary conditions from the *T(ξ,y,z)* function to the *φ(ξ,y,z)* function, we can easily solve this equation with a standard Lagrangian *FEM* code that contains a solver for positive definite symmetric matrix. Equation (3-7) applies only to linear systems (i.e. *α* is independent of *T*) because of equation (3-6).

3.3 Weld Heat Source

What we know about a weld heat source either comes from experimental observation or more detailed models of the welding process. Experimentally, currents, voltages, frequency, wire feed rates, welding speeds, etc. can be measured. Also welds can be sectioned and cross-sections of a weld measured optically. Various forms of video cameras, some with laser illumination, can be used to measure the weld pool surface and to visualize droplet transfer into the weld pool. Thermocouples and infra-red cameras can be used to measure temperatures in and near the weld pool. The distribution of power density in arcs, lasers and electron beams can be measured. Spectroscopy of the welding arc plasma provides useful information. A great deal of experimental data can be obtained for any given weld and a great deal of data of this type has been published.

The other source of knowledge and data about welding heat sources comes from mathematical models of the weld heat source. Such models can include in addition to the energy equation, surface tension of the weld pool surface, hydrostatic forces, Lorentz and Marangoni forces in the weld pool, pressure and shear forces from the arc or plasma, and droplet transfer to the weld pool. Once we are able to make models that predict the behavior we observe, we tend to say that we understand the welding process. It is equally important to note that to the extent that we cannot make models that predict the behavior of a welding process, we should say that we do not understand the welding process.

3.3.1 Data to characterize a Weld Heat Source

The best way of modeling a weld heat source depends on many factors.

The first factor to consider is how accurately we want to model the heat source. Few if any welding processes used in industry are controlled more accurately than 5% and many are controlled less accurately. Our knowledge of the values of material properties such as thermal conductivity, specific heat, latent heats, Young's modulus and Poisson's ratio, etc. are rarely known with accuracy greater than 5%.This immediately restricts the accuracy that can be achieved by a model to not more than say 5 to 25 %.

The second factor is our objective in modeling. What do we want to use the weld heat source model for? If it is to be used to predict hot cracking, then it will have to be accurate near the weld pool and point heat sources models will be of little use. If the purpose of the weld heat source is to predict distortion and residual stress in low alloy steel structures, then accurate temperatures below about *600* to *800°C* could be most important. Temperatures above this range have much less effect on distortion and residual stress.

The third factor is what information is available for use in a weld heat source model. Perhaps the simplest models use only the weld power input and the shape and size of the weld pool. Important data can be obtained from a macrograph of a weld cross-section. A useful estimate of the front weld pool is a circular arc with diameter equal to the width. The length of the back half of the weld pool is often of the order of twice the width of the weld pool. These rather rough estimates of the weld pool size and shape are often useful when more accurate data describing the weld heat source are not available. These data are relatively easy and cheap to obtain. These data tell us little or nothing about the physics inside the weld pool and in the weld arc.

Models that include the magneto-hydrodynamics of the arc, fluid flow in the weld pool, can involve unstable phenomena such as turbulence that lead to mathematical problems that can be extremely difficult to solve if they can be solved at all. Except in low power welds where the instabilities tend not to arise, the value of such models has been primarily in improving our understanding of the importance of various terms such as Lorentz force, buoyancy force, surface tension forces, etc. In the opinion of these authors, these models have not yet been useful for predicting the behavior of high power production welds.

In these authors' opinion, the most accurate models of weld heat sources available today are those developed by Sudnik [24, 25, 26 and 27] and his colleagues for specific welding processes. They parameterize the welding process and the parameters are chosen to capture the most important physics of the weld pool. In addition the parameters are carefully correlated with experiment for a given welding process. Although the time and money needed to develop these models for each welding process can be considerable, they have the very important advantage of being the most accurate predictive weld heat source models available. Once developed, they should last for the life of the welding process. In this sense, developing such a weld heat source model is a one-time expense.

3.3.2 Modeling a Weld Heat Source

Having specified the data available to characterize the weld heat source, the next step is to decide how to use the available data to compute the transient temperature field of the weld in the structure being welded. The most popular approaches are listed below. They are roughly in chronological order because the more recent generations have usually extended previous generations.

Most welding processes use a heat source such as an arc, plasma torch, laser or electron beam. The temperature field due to heat from the heat source melts the base metal creating a weld pool. In addition, filler metal from an electrode is often added to the weld pool. The weld pool moves along a weld joint and as the weld pool solidifies, it creates one pass of a weld joint. The temperature field driven by the weld heat source is the dominant driving force of the welding process. It causes phase transformations, thermal strain and thermal stress, distortion and residual stress. To analyze or predict the behavior of a weld in a structure, this transient temperature field must be computed with useful accuracy. The transient temperature outside of the weld pool depends primarily on the distribution of energy from the heat source and the conduction of heat away from the weld pool by conduction in the solid. Stress and strain usually have little effect on the transient temperature field.

We will assume that the heat source is an arc unless otherwise stated. However, almost everything we say would apply equally well to most other weld heat sources.

Because the physics of the weld heat source are often complex, thermal models of weld heat sources have been developed to reduce the complexity to more manageable levels. These models allow the heat equation or energy equation to be solved for an approximate solution while ignoring most of the physics of the welding process. Except in the interior of the weld pool itself, these approximate solutions can be remarkably accurate if they are given good data. If the physical phenomena in the interior of the weld pool are of interest, then a solution of such a thermal model is usually the starting point for solving the equations that model other physical phenomena in the weld pool, such as stirring of the liquid metal.

We will classify weld heat source models into categories of First Generation to Fifth Generation. First Generation is the oldest and simplest and Fifth Generation is the newest and most complex. Each older generation is a sub-class of every newer generation. By removing features from a younger generation one moves to an older generation. By adding features to an older generation, one moves from an older generation to a younger generation.

First Generation Weld Heat Source Models

First Generation weld heat source models are the point, line and plane heat source models of Rosenthal [18] and Rykalin [17]. They are the first and most famous of weld heat source models. The point source specifies the position of the weld and the net total rate of heat input from the weld source *(J/s)*. This model is useful for shallow weld pools on thick plates. The line heat source specifies a line segment and the net power or net rate of heat input per unit length uniformly distributed along the line segment. This line weld heat source model can be useful for full penetration laser and electron beam welds in sheets and plates. The sheet weld heat source model can be useful for overlay welds made with sheet electrodes on very thick plates.

The point weld heat source models can be quite accurate when they are used to evaluate temperatures sufficiently far from the weld pool. Concentrating the energy in a welding process into a mathematical point can be viewed as simply a mathematical trick to avoid dealing with the real distribution of energy in a weld heat source. If the power in a weld heat source was actually contained in a point, then the power density and the temperature would be infinite which is impossible. In spite of this defect, the value of these models should not be under estimated. They conserve energy which any useful model of a weld heat source must do and they provide quite accurate temperature distributions at distances sufficiently far from the weld. This is a significant accomplishment.

It should also be recognized that the fundamental strength and fundamental weakness to these point, line and plane heat sources is that they ignore and thus they fail to account for the distribution of energy in the real weld heat source. Any significant improvement on these heat sources must account more accurately for the distribution of energy in the weld heat source.

There are other criticisms of these weld heat source models. For example, the analytic solutions based on these models are linear and thus assume temperature independent material properties, and do not deal with phase transformations. Because they are steady state models, they cannot deal with starting and stopping transients. Also they require the weld to be in a prismatic body and the weld to travel parallel to the axis of the prism. Including convection and radiation boundary conditions is awkward. These authors regard these deficiencies as minor in comparison with the need to distribute energy realistically in the weld heat source. If the power density distribution in the weld pool was realistic, most of the other limitations could be handled.

In summary, these models act on very simple geometric domains such as infinite sheets or plates. They only specify the weld position with a straight weld path as a function of time and the total power input. A spiral path on an infinitely long cylinder is also possible. They only determine steady state solutions of the temperature field.

The weld pool is not an essential part of these models. One can of course define the weld pool to be the region with temperature greater than the melting temperature. However, the model itself contains no notion of latent heats or phase transformations.

Second Generation Weld Heat Source Models

Second Generation weld heat source models replace the point, line and plane models that mathematically are delta functions, with distribution functions. The first of these was a distributed flux model by Pavelic [19] and by Rykalin [17], see Chapter *II*. This distributed flux model was particularly effective for low power density welds in which the weld pool does not have a nail head or other form of deep penetration. However it could not model deep penetration welds such as electron beam, laser or plasma welds and most high power arc welds such as those with a nail head cross-section. Goldak et al [1] proposed a distributed power density model that could model deep penetration welds with somewhat more complex weld pool shapes. The last of these Second Generation models is the prescribed temperature model proposed by Goldak et al [2 and 14], see sections 3-3-2-2-2 and 3-4. It could easily model weld pool shapes of arbitrary complexity. Later we will see that if the only equation to be solved is the heat equation, then the only functions that could be distributed in a weld heat source model are the power density, a prescribed flux and a prescribed temperature. Of course one might choose various distribution functions and one could pulse or weave the heat source. However, there can be no other classes of functions for Second Generation weld heat source models because we define them to be based solely on the heat equation.

Second Generation weld heat sources immediately remove most of the limitations of the First Generation models. They can model domains of complex geometry. Nonlinearities such as temperature dependent thermal conductivity, specific heat, radiation and convection boundary conditions, latent heats of phase transformation and microstructure evolution are easily included or coupled. These models can have but need not have a weld pool. If the region being modeled is defined to be just outside the weld pool, then any temperature distribution in the region that would be occupied by the real weld pool is fictitious. Even if these models include the weld pool, the temperature distribution in the weld pool and the shape of the weld pool are often crude approximations of reality. Since these models have no velocity distribution in the weld pool, this contributes an additional error in any computed temperature field computed in the weld pool by these models. Thus these models ignore most of the physics of the weld pool. Nevertheless, the weld pool shape and position data can be realistic, because these models can compute accurate temperature distributions outside of the weld pool.

Distributed Heat Source Models

The first second Generation models define a distributed heat source function. The best known example of such a function is the double ellipsoid model, see Figure 2-9. Another example is a conical model for deep penetration electron or laser beam welds, see Figure 2-11. These were the first weld heat source model capable of simulating welds with deep penetration. Linear combinations of such ellipsoidal, conical or other shapes can be used. If one knows the "exact" temperature solution, then one can substitute it in the energy equation to determine the "exact" distributed heat source function directly. Thus the distributed heat source function exists whenever the temperature solution in the weld pool region exists. However for weld pools with complex shapes, it can be very difficult to determine the distributed heat source function if the temperature distribution in the weld pool is not known.

Because of the discrete *FEM* mesh, one should constrain the net heat input of the weld heat source model to equal the net heat input of the weld. This is easily done by scaling the load vector for the heat source model.

The *FEM* mesh can conform to the weld pool boundary but it need not conform.

Prescribed Temperature Heat Source Models

These models treat the weld heat source as a sub-domain in which the temperature or specific enthalpy in the weld pool is known as a function of (x,y,z,t). Since the temperature is known, it need not be solved. Instead, the boundary of this sub-domain can act as a Dirichlet *BC* for the complement of the weld heat source sub-domain. Perhaps the simplest useful example of this class is parameterized by a double ellipsoid function that specifies the liquidus temperature on the liquid solid interface and a maximum temperature at the centroid of the weld pool and constrains the temperature to vary quadratically from the melting point to the maximum temperature.

Radaj [4 and 8] proposed a model based on the liquidus surface and liquidus temperature. For the continuous problem, i.e., before the problem is discretized for *FEM* or *FDM*, this interface partitions the domain into the weld heat source model and its complement. However, after discretization, the problem arises of how to map this surface into the discrete problem. If the weld pool is meshed so that the boundary of the weld pool lies on faces of elements and if the mesh moves with the weld pool, then a surface representation of a weld pool has distinct advantages. If the weld pool moves through the mesh in time, then the advantages of representing a weld pool by the liquid-solid surface diminish.

Sudnik [20] uses a *FDM* with a regular grid to solve for specific enthalpy in his weld heat source models. He chooses a rectangular Cartesian grid with edge lengths $P*dx$, $Q*dy$, and $R*dz$ where P, Q and R are integers and dx, dy and dz are the lengths of edges of each cell. These models could be used directly in a *FDM* or *FEM* method for solving the transient temperature in a structure being welded. There is no need to restrict the weld heat source model to the weld pool. The farther the weld heat source model can be extended from the weld pool, the coarser the mesh required by the structure being welded and the lower the computing costs. In particular, ideas from A. Brandt's multi-grid method can be used which would treat Sudnik's model as a local fine grid solution. The thermal analysis of the structure could use a coarse grid with the fine grid solution in the weld pool region restricted to the coarse grid.

Third Generation Weld Heat Source Models

The next advance in weld heat source models, Third Generation models, was initiated by Ohji et al. [21]. They predicted the liquid weld pool shape. The distinguishing feature of Third Generation models is that they must solve the Stefan problem for the weld pool liquid-solid free boundary. Recall that First and Second Generation models need not and usually do not solve the weld pool liquid-solid free boundary problem. In these Third Generation models the specific enthalpy is discontinuous across the liquid-solid interface. The melting temperature is usually defined for a flat static interface. Increasing curvature decreases the melting point. Interface velocity increases the melting point in the front facing part of the weld pool and decreases it in the backward facing part of the weld pool.

In addition to the energy equation with liquid-solid free boundary, these models include the hydrostatic stress in the liquid pool, a pressure distribution on the weld pool surface from the arc, surface tension forces, and a volume constraint to enforce conservation of mass, i.e., the mass entering the weld pool equals the mass leaving the weld pool. This force balance is the conservation of linear momentum with velocity set to zero or the momentum equation for a hydrostatic fluid. Thus this model couples the heat equation with the momentum equation for a hydrostatic fluid. Given the input data, these models can predict the weld pool shape and the shape of the weld bead created by the weld pool. Sudnik has developed these weld heat source models to the current state of the art. Weiss [22] extended Sudnik's ideas to include some effects of the arc interaction with the weld pool shape. Weiss was able to predict effects of vertical, horizontal welding on weld pool shape in addition to flat welding.

The Third Generation Models ignore the Lorenz force, the Marangoni force and the force due to the momentum flux from any droplets added to the weld pool from a consumable electrode. Except for the model by Weiss, they usually assume that the flux and pressure distribution from the arc are constant and independent of the weld pool free surface shape. Numerically, these models are robust and computational costs are only slightly higher than First or Second Generation Models. While the data required by the First and Second Generation Models was the distribution of power density, flux or temperature, the data required by these Third Generation Models usually include a pressure distribution from the arc, a mass flow rate into the weld pool and surface tension on the liquid surface of the weld pool. Now the geometry of the weld pool is not input data but output data.

Fourth Generation Weld Heat Source Models

Fourth Generation models are distinguished by adding the equations of fluid dynamics to the modeling of the weld heat source.

Recall that the First, Second and Third Generation models have no fluid velocity. The most general equations for macroscopic fluid dynamics are the Navier-Stokes equations. They can include buoyancy and Lorentz forces acting on the interior of the liquid phase. Marangoni effect forces, pressure and shear forces from the arc act on the surface of the weld pool. Some models include some form of droplet flow from a consumable electrode. However, as arc welding currents rise much above *100* to *150* amps, most of these models have difficulty accounting for what appears to be chaotic motion in the weld pool and possibly chaotic and possibly turbulent motion coupling the velocity, pressure and temperature of the liquid in the weld pool and velocity, pressure, temperature, voltage and current density in the arc. To the author's knowledge, the current state of the art of Fourth Generation models that couple heat equation with the transient Navier-Stokes equations has not pushed past this barrier. On the other hand, Sudnik [25 and 26] has developed models for deep penetration laser welds that contain a few, say three, functions describing the velocity field in a laser weld pool. He calibrates these models with experimental observations to estimate a few coefficients or fitting parameters in the model. These models are very robust and require very little computing power or time. Because they are correlated with experimental data they are expected to be accurate for the ranges of welds for which they were fitted. The price to be paid is the cost of the experiments needed to determine the correlation coefficients.

Fifth Generation Weld Heat Source Models

Fifth Generation models make a serious effort to include a model of the arc in the heat source model. This adds the equations of magneto- hydrodynamics to the equations in the previous models. Although these Fifth Generation models are very general, they face even more serious mathematical difficulties than the Fourth Generation Weld Heat Source Models. Often existence and uniqueness cannot be proven. The numerical methods face very serious difficulties.

This is not to argue that computer simulations of Fourth and Fifth Generation models are not useful. It is to argue that to date *(2003)* their usefulness for predicting the geometry of the weld pool in high power industrial welds has been limited.

Hierarchical Weld Process Models

When analyzing a given weld, one is not required to remain faithful to a single model. Clearly, unless one wishes to analyze the interior of the weld pool, it is sufficient for all analyzes outside of the weld pool to specify the geometry of the weld pool as a trimmed surface patch preferably in the solid. Then for each point in space-time on this surface patch specify the temperature, composition, phase fractions and tractions. All coupling in the welded structure with the welding process can be controlled by this interface.

We prefer the interface to be in the solid in order to leave all issues related to the mushy zone in the weld pool model. In most cases, the interface itself could be represented by four cubic triangular patches or *FEM* elements. This would need *25* nodes or points and usually would be more accurate than the data available to characterize the welding process.

The weld pool itself could be represented in most cases by four cubic tetrahedrons. This has the advantage that the surface is the boundary of this weld pool and all data is defined every where in the weld pool and not just on the surface of the weld pool. It could be advantageous to add a layer of cubic bricks in the solid to provide an overlap with the far field domain.

We recommend that whatever model is used to define the weld pool data, the Second Generation prescribed temperature model be used for all analyzes outside of the weld pool. This strategy simplifies the code and increases computational efficiency.

3.4 Heat Transfer in Welds

The most popular model for the heat input is double ellipsoid, because for many arc welds the double ellipsoid shape is a good approximation. It shows that a Gaussian distribution of power density inside a double ellipsoid moving along the weld path was convenient, accurate and efficient for most realistic welds with simple shapes. Since the model reflects the depth and shapes of a weld, it has proved to be more realistic and flexible in application than previous models.

However, the power density model must generate the correct shape of the weld pool. Unfortunately, when the shape of a weld pool is more complicated than the shape of a double ellipsoid, e.g. the double pool in a submerged arc weld described by Barlow [5], Figure 3-2, it is very difficult to find a power density distribution function that accurately computes the transient temperature field.

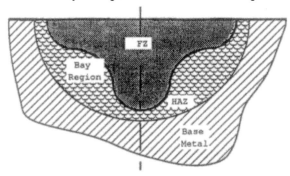

Figure 3-2: Cross-section of the deep penetration Barlow weld containing a "bay" region.

The determination of the temperature field only by the melting point temperature at the liquid-solid interface is defined by Radaj as the equivalent heat source (equivalent, because the temperature distribution in the weld pool replaces the power density distribution). For this purpose Goldak used the term 'prescribed temperature heat source'.

The prescribed temperature distribution function has been developed to model weld pools with more complex geometry because:

-The temperature distribution can be measured more accurately through experiment [6] than the power density distribution,

-At the liquid-solid interface the temperature can be set to the melting point of the base metal, possibly adjusted for curvature and interface speed,

-The cross-section of the prescribed volume can be measured from a metallograph as Barlow [5] and Glickstein and Friedman [7] have demonstrated,

-The effects of radiation and convection on the surface of the weld pool are already incorporated in the prescribed temperature field and need not be included in the weld model,

-The computing time is reduced because the prescribed degrees of freedom need not be evaluated.

The temperature field also can be obtained by specifying the heat input as described above or by prescribing the temperature in the weld pool. Gu et al [2] presented a model with prescribed temperature distribution in the weld pool, which demonstrates that the prescribed temperature field can model complex weld pools accurately; see section 3-4-5-1.

A finite element calculation based on an Eulerian formulation for steady state temperature fields for welds with filler metal addition derived from the transformation eq. (3-6) is also presented by Gu et al. in [14], in which the thermal diffusivity is taken as a constant, see section 3-4-5-2.

3.4.1 Power Input

The net power input for the weld, volt × amp × efficiency/speed, should equal the total thermal load reaction or sum of the Lagrange multipliers at the nodes connected to the prescribed temperature field in the weld pool. This is a consistency check and a redundant measure of correctness.

Since the total heat input is determined only by the temperature field at the liquid-solid boundary, i.e., $q = -\kappa\,T_n$, the net power input can be achieved by adjusting the weld pool size. The power input does not depend on the temperature of the interior of the weld pool even if the weld pool is a cavity. In a weld with a double pool, the metal in the leading pool is usually blown away in a real weld. In the model, the absence of molten metal in the cavity can be ignored without penalty.

If the region near the arc is studied in detail, then the double ellipsoid model for prescribing the heat input may be better [3]. This should then be combined with a fine mesh with about 10 linear elements across the axis of the ellipsoid area of heat input [3 and 8]. The heat input can be adjusted to fit the experimental data. This can be for example the fusion zone, but the highest accuracy can be obtained when transient temperatures have been measured.

Jones et al. [9 and 10] prescribed the temperature at the boundary of the weld pool which was defined before hand by the user to be equal to the $T_{liquidus}$. They simulated the welding of a bead on a disc and compared temperatures with measured temperatures and obtained quite good agreement. Figure 3-3 compares measured temperatures at some different locations with computed values.

It is not difficult to achieve good accuracy for the temperatures at some distance from the weld, especially if only a few weld passes are made [3]. The problem of matching measured temperatures is naturally greater when the measurement is performed closer to the arc. Figure 3-4 shows that the agreement is quite good at weld 19 when the thermocouple is farther away from the arc but not so good during later weld passes when the arc comes closer to the thermocouple.

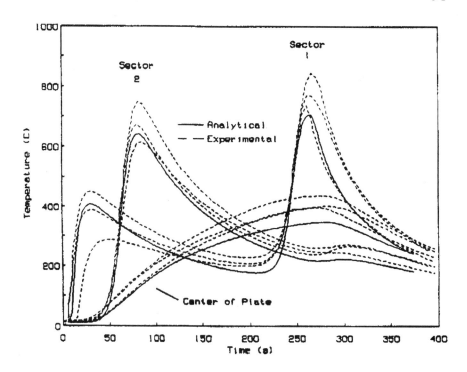

Figure 3-3: Computed and measured temperatures for the bead on a disc, from Jones et al. [10].

Roelens [11 and 12] and Lindgren et al. [3 and 13] prescribed the temperature in the case of multi pass welds. The temperature for a weld pass is specified during the heat input phase, as shown in Figure 3-5. Results from this technique for a 28-multipass weld are shown in Figure 3-4.

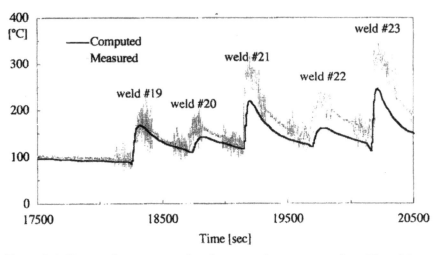

Figure 3-4- Excerpt from computed and measured temperature for a 28-multipass weld, from Lindgren [3 and 13].

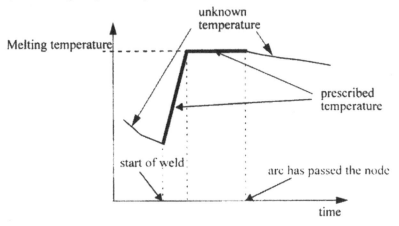

Figure 3-5: Prescribed temperature for node associated with a weld, from [3]

The temperature levels in the neighborhood of the peak temperature may be more important to match than the peak itself. Then the correct total amount of net heat input is obtained. Since the weld pool has dimensions of the order of a cm and the structure usually has dimensions of the order of *10s* of *cm*, the mesh must be graded to reduce computational costs to affordable levels.

3.4.2 Implementation of prescribed temperature model

The welding path is assumed as any valid geometric curve in space on which a local coordinate system moves. From the local coordinate system, a tangential vector and principal normal can be computed at any point on the curve.

Typically, a point is calculated according to the velocity and the time as the leading weld pool travels on the path. Such a point can serve as the centre of the leading weld pool. The locations of other weld pools on the path are specified relative to the centre of the leading pool.

At the centre of each weld pool, an outward normal is calculated. This can be computed as the average of the outward normal of all free surfaces of the elements containing the centre.

For each weld pool, unit vectors for the outward normal and tangential direction define a local coordinate system. The distribution function in a weld pool has a standard form, e.g., a double ellipsoid. The local coordinate system consists of (x^1, y^1, z^1) in this coordinate system. The transformation matrix from global to local coordinates is:

$$[H] = \begin{bmatrix} x^1 \\ y^1 \\ z^1 \end{bmatrix} = \begin{bmatrix} x_x^1 & x_y^1 & x_z^1 \\ y_x^1 & y_y^1 & y_z^1 \\ z_x^1 & z_y^1 & z_z^1 \end{bmatrix} \tag{3-8}$$

To identify the small subset of nodes in the vicinity of the weld pool from the total set of nodes in the mesh, which usually are enormous, a *KDTree* provides an *O(log N)* algorithm.

A transformation from global to local coordinates can be done on the set of nodes selected by the *KDTree*.

$$\begin{bmatrix} x^1 \\ y^1 \\ z^1 \end{bmatrix} = [H] \begin{bmatrix} x - x_0 \\ y - y_0 \\ z - z_0 \end{bmatrix} \tag{3-9}$$

where x, y and z are global coordinates and x_o, y_o and z_o define the origin of a local coordinate system (x^1, y^1, z^1).

For any distribution function, including a double ellipsoid, all nodes can then be tested to determine if they are in the domain of the distribution function. For example, a node is in the front part of an ellipsoid if and only if:

$$x^1 > 0 \qquad (3\text{-}10)$$

and

$$\frac{(x^1)^2}{a_1^2} + \frac{(y^1)^2}{b^2} + \frac{(z^1)^2}{c^2} \leq 1.0 \qquad (3\text{-}11)$$

If this is true, the node is prescribed with a temperature:

$$T = T_0 \exp\left\{ A\left(\frac{(x^1)^2}{a_1^2} + \frac{(y^1)^2}{b^2} + \frac{(z^1)^2}{c^2} \right) \right\} \qquad (3\text{-}12)$$

where T_0 is the temperature at the origin, which has the maximum temperature, and A is a constant which has to be evaluated so that $T=T_m$, the melting point temperature at the boundary of *FZ* and *HAZ*.
If:

$$\frac{(x^1)^2}{a_1^2} + \frac{(y^1)^2}{b^2} + \frac{(z^1)^2}{c^2} = 1.0 \qquad (3\text{-}13)$$

and

$$T_m = T_0\, e^A \qquad (3\text{-}14)$$
$$A = ln\ (T_m/T_0)$$

Substituting A into equation (3-12);

$$T = T_0 \left(\frac{T_m}{T_0} \right) \qquad (3\text{-}15)$$

The nodes inside any distribution function can be identified and assigned temperatures in this way.

3.4.3 Starting Transient

When a real weld is started, the weld pool is not formed instantly. First, a Gaussian flux distribution heats the solid to the melting point, then a thin layer of molten metal forms. This grows in width and depth with time. With increasing depth, the weld pool surface is depressed and stirring is induced by electro-magnetic, buoyancy, arc pressure and surface tension gradient forces.

Modeling this starting transient is much more complex than modeling the steady-state weld pool. A simple technique is to hold the weld pool stationary for a few seconds. In practice, welders usually do this. The total energy input for the prescribed temperature heat source for time *(0, t₁)* will equal the energy input for a constant power heat source for a time *(0, t₂)*, where *(t₁<t₂)*. Using the shorter time, t_1, with the prescribed temperature heat source reduces this error.

There is also a second order error in the starting transient temperature field because of time-dependent diffusion.

During the starting transient of the weld, the natural boundary condition models, i.e., thermal flux models develop the weld pool more realistically.

Current technology uses distribution functions with constant size and shape. It would be more realistic to interpolate the distribution function from a Gaussian flux distribution on starting to the steady state flux and power density distributions. The prescribed temperature distribution function is quite unrealistic on starting. In terms of an electrical circuit analogy, it applies a constant voltage for charging a capacitor. This results in a high power starting transient. However, the effect on the temperature field at later times is small and unless one is specifically interested in the starting transient, it can be neglected. It has been proposed that this problem could be resolved by applying the prescribed temperature heat source until the region near the weld pool reaches steady state. At that time compute the power density and flux distribution. Then restart the weld with this power density and flux distribution.

3.4.4 Boundary Conditions

On the boundary of the domain Ω either the essential (prescribed temperature) or natural (prescribed flux) boundary conditions must be satisfied. The essential boundary condition can be defined as:

$T(x,y,z,t) = T_1(x,y,z,t)$ on the boundary S_1; i.e., $(x,y,z) \in S_1: t > 0$ (3-16)

The natural boundary condition can be defined as:

$$\kappa_n \frac{\partial T}{\partial n} + q + h(T - T_\circ) + \sigma\varepsilon(T^4 - T_\circ^4) = 0 \qquad (3\text{-}17)$$

On the boundary S_2: i.e. *(x,y,z)* $\in S_2$: *t* > 0

κ_n = thermal conductivity normal to the surface *(W/mC)*

q(x,y,z,t) = a prescribed flux *(W/m²)*

h   = heat transfer coefficient for convection *(W/m²C)*

σ   = Stefan-Boltzmann constant *(W/m²C⁴)*

ε   = emissivity

T_0 = the ambient temperature for convection and/or radiation *(C)*

If radiation is included or if the convective heat transfer coefficient is temperature dependent this boundary condition is nonlinear.

In addition, the initial condition must be specified for *(x,y,z)* $\in \Omega$:
$$T(x, y,z,0) = T_0\,(x,y,z) \qquad (3\text{-}18)$$

If the partial differential equation (3-1), the boundary conditions (3-16) and (3-17), and the initial condition (3-18) are consistent, the problem is well posed and a unique solution exists.

The most convenient domains in which to apply equations (3-7) are prisms with constant cross-sections, as found in the example shown in Figure 3-6. The temperature of the boundary between the weld pool and the solid is prescribed to the solidus temperature. By using an essential boundary condition, Gu et.al. [2 and 14], the heat entering the base metal through the interface, Γ_3, equals the thermal reactions, and the flux due to complex radiation and convection on the surface of the weld pool need not be taken into account.

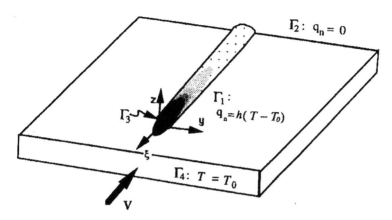

Figure 3-6: The Eulerian frame and boundary conditions. Γ_2 is the downstream boundary that is normal to the velocity of the source. On this surface the gradient of the temperature has been prescribed to zero. Γ_3 is the surface of the liquid-solid interface of the weld pool. On this surface, which is smooth but with arbitrary orientation and curvature, the temperature has been prescribed. Γ_4 is the surface of the upstream end. The outward normal of the surface is parallel to the velocity of the source. On this surface the ambient temperature of the plate has been prescribed. Γ_1 is the remainder of the boundary surface. The outward normal of the surface is orthogonal to the velocity of the source. On this surface natural convection has been applied.

From experimental measurements, we know that at a sufficient distance, ξ_n, in front of the weld pool, temperatures remain at ambient temperature; $T=T_0$ when $\xi > \xi_n$ and consequently, $\varphi=0$ when $\xi > \xi_n$.

When the prism boundaries as Γ_1 are not insulated, a boundary flux would have to be applied. For the case of a constant convective coefficient, h, the boundary conditions is:

$$q_n = -k\nabla T \cdot n_\Gamma = h(T - T_0) \qquad (3\text{-}19)$$

where q_n is the scalar external flux and n_Γ is the outward normal vector on the boundary.

We express equation (3-19) in terms of φ, using equation (3-6):

$$\kappa\nabla T \cdot n_\Gamma = \kappa e^{-(\alpha v \xi)/2}\left(\nabla\varphi \cdot n_\Gamma - \frac{\alpha v}{2}\varphi n_\xi\right) = -he^{-(\alpha v \xi)/2}\varphi \qquad (3\text{-}20)$$

On the side walls of the prism, n_Γ is orthogonal to the weld path,

$$n_\Gamma = \begin{bmatrix} 0 \\ n_y \\ n_z \end{bmatrix} \qquad (3\text{-}21)$$

Equation (3-20) for the side walls becomes:

$$\nabla\varphi \cdot n_\Gamma = -\frac{h}{k}\varphi \qquad (3\text{-}22)$$

The heat source term has been set to zero. If one wished to include the weld pool, a heat source term exists that could replace the prescribed temperature boundary condition on the weld pool solid-liquid interface.

If the domain is long enough or infinite, at the downstream end the temperature will drop to ambient temperature because of the heat dissipation from the surfaces. Either essential or natural boundary conditions can be applied, i.e., $T=T_0$ or $\nabla T \cdot n_\Gamma = 0$ when $\xi \Rightarrow -\infty$. Transforming temperature T to the potential function φ, when $\nabla T \cdot n_\Gamma = 0$ is applied and $\xi \Rightarrow -\infty$,

$$\nabla T \cdot n_\Gamma = e^{-(\alpha v \xi)/2}\left(\nabla\varphi \cdot n_\Gamma - \frac{\alpha v}{2}\varphi n_\xi\right) = 0 \qquad (3\text{-}23)$$

So,

$$\nabla\varphi \cdot n_\Gamma = \frac{\alpha v}{2}\varphi n_\xi \qquad (3\text{-}24)$$

3.4.5 Finite Element Solutions with Prescribed Temperature

Of the three strong candidate numerical methods; finite difference, boundary element and finite element analysis, the finite element method have been chosen for its capability for nonlinear analysis and dealing with complex geometry. In addition, it is most compatible with modern *CAD/CAM* software systems. For thermal analysis alone, a strong argument can be made in favour of finite difference methods. However, for thermoplastic analysis the argument is stronger for finite element analysis. The boundary element method is not well developed for nonlinear analysis. Briefly, these considerations led to the choice of *FEA* as the most effective numerical method for developing a complete analysis capability for computer modeling or simulation of welds.

The finite element method *FEM* usually imposes a piece-wise polynomial approximation of the temperature field within each element:

$$T(x,y,z,t) = \sum_{i=1}^{nodes} N_i(x,y,z)T_i(t) \qquad (3\text{-}25)$$

where N_i are basis functions dependent only on the type of element and its size and shape. Physically T_i (t) is the value of the temperature at node i at time t.

$$\left[\frac{\partial T}{\partial x}, \frac{\partial T}{\partial y}, \frac{\partial T}{\partial Z}\right] = \left[\frac{\partial N_i}{\partial x}T_i, \frac{\partial N_i}{\partial y}T_i, \frac{\partial N_i}{\partial z}T_i\right] \qquad (3\text{-}26)$$

where $\sum N_i(x,y,z)T_i(t)$ is abbreviated to N_i T_i.

The next question is how to evaluate T_i? Galerkin's *FEM* is among the most convenient and general of the methods available for this purpose. If eq. (3-25) is substituted into eq. (3-2), a residual or error term must be added. If this was not true eq. (3-25) would be the exact solution. Indeed when eq. (3-25) is the exact solution, the error in the *FEM* solution is zero.

Galerkin's *FEM* method requires:

$$\int_\Omega \varepsilon N_i \, d\Omega = 0 \qquad (3\text{-}27)$$

Mathematically ε is the residual from eq. (3-2); N_i in eq. (3-27) is a test function and the N_i terms in eq. (3-25) are the trial functions. Since there are i nodes, eq. (3-27) creates a set of i ordinary differential equations which are integrated to form a set of algebraic equations for each time step:

$$[K]\,[T] = [R] \qquad (3\text{-}28)$$

This set of equations is solved for the nodal temperatures T_i at the end of the time step. Usually some form of Newton-Raphson method together with a Gaussian elimination and back substitution would be employed.

Applying Galerkin's method to the equation (3-7), when $\tilde{Q}$ is zero gives:

$$\int_\Omega W\left(\nabla^2\varphi-\frac{(\alpha v)^2}{4}\varphi\right)d\Omega=0 \qquad (3\text{-}29)$$

for all W in the space of test functions.

For infinite domains the test functions must be specified carefully but for a finite domain the usual Galerkin test functions can be used. Apply Green's identity to equation (3-29):

$$\int_\Omega\left[-\nabla W\nabla\varphi-\frac{(\alpha v)^2}{4}W\varphi\right]d\Omega+\int_{\partial\Omega}W(\nabla\varphi\cdot n_\Gamma)d\Gamma=0 \qquad (3\text{-}30)$$

Applying the boundary condition in equation (3-22) to the side walls,

$$\int_{\Gamma_1}W(\nabla\varphi\cdot n_\Gamma)d\Gamma=-\int_{\Gamma_1}\frac{h}{k}W\varphi d\Gamma \qquad (3\text{-}31)$$

If the boundary condition in equation (3-24) is applied on the downstream end,

$$\int_{\Gamma_2}W(\nabla\varphi\cdot n_\Gamma)d\Gamma=\int_{\Gamma_2}\frac{\alpha v}{2}W\varphi n_\xi d\Gamma. \qquad (3\text{-}32)$$

When the end faces are perpendicular to ξ, $n_\xi = -1$ at the rear end.

Putting equations (3-31) and (3-33) into (3-30):

$$\int_\Omega\left[-\nabla W\nabla\varphi-\frac{(\alpha v)^2}{4}W\varphi\right]d\Omega-\int_{\Gamma_1}\frac{h}{k}W\varphi d\Gamma-\int_{\Gamma_2}\frac{\alpha v}{2}W\varphi d\Gamma=0 \quad (3\text{-}34)$$

for all test functions W.

3.4.6 Computational Results

Complex Weld Pool Shape

Computational results are presented for the experiments described by Barlow [5]. It is a weld deposited on a low carbon steel plate of thickness *19 mm*. The weld length is about *250 mm*. In order to minimize end effects, the length and width of the plate are chosen to be *1224 mm* and *500 mm*, respectively. This is consistent with

Barlow's experiment where the plate is considered infinite in length and width.

The temperature-dependent thermal conductivity and volumetric heat capacity are taken from Reference [15].

The total number of elements is *12092* and the total number of nodes is *15323*. Most elements are concentrated along the weld path for high resolution in the *HAZ*, where the temperature gradient changes rapidly during welding.

Several levels of grading elements are used [16] to allow a coarse mesh in the far weld area. This reduces greatly the computing cost without losing significant accuracy.

The size of the weld pool, see Figure 3-7, was determined from the experimental results in Reference [5]. The same weld parameters as condition 1, Reference [5], were applied in order to compare the computed results with the results of experiment. The nominal power input is *3.2 kJ/mm* and the weld speed is *6.67 mm/s*.

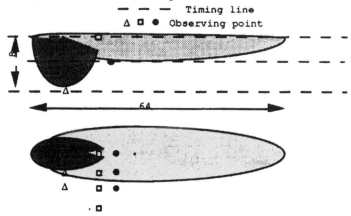

Figure 3-7: Frame of weld pool, the compound ellipsoid model. The temperatures at nodes inside the pool are prescribed, from [2].

As mentioned before, the prescribed temperature model used as the heat source is correct only in the steady state, i.e., when the heat input to the weld pool does not change with time. The results presented here are those at about *24 s* from the beginning.

According to the node reaction to power input, Figure 3-8, it can be seen that a relatively steady state is formed as early as *5 s* after

striking the arc. Since the electrode is usually held still for a few
seconds after the arc is struck, the total energy input per unit length
(kJ/mm) is greater than the nominal value in the steady state. This
total energy input is matched by adjusting the length of time the
prescribed temperature weld pool is held stationary.

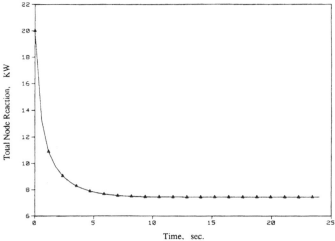

Figure 3-8: Decay of power input vs. time for prescribed temperature weld pool
model for the Barlow weld. Power input is the summation of node reactions, from
[2].

Figure 3-9 shows the contours of the temperature field. The
region near the weld pool is magnified in Figure 3-10. The deep
leading pool directly under the welding arc can be seen. It is
suspected to be an empty hole in the real weld because the liquid
metal is blown out into the shallower rear pool. The rear pool is
formed to accumulate this ejected liquid metal. It is shallow and
wider, so that the temperature gradient is relatively small within the
rear pool. Between the front and the rear pools, there is only a thin
layer of molten metal.

The temperature contours outside the *HAZ* are approximately
elliptical in shape. A similar result could be obtained using a single
weld pool, Figure 3-15.

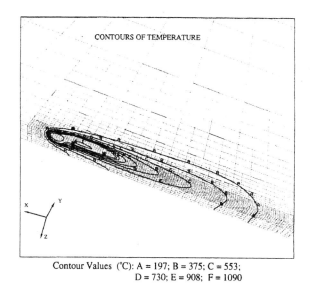

Contour Values (°C): A = 197; B = 375; C = 553;
D = 730; E = 908; F = 1090

Figure 3-9: The temperature field 24 s after striking the arc. The minimum temperature = 20.0°C; the maximum temperature = 1800°C, from [2].

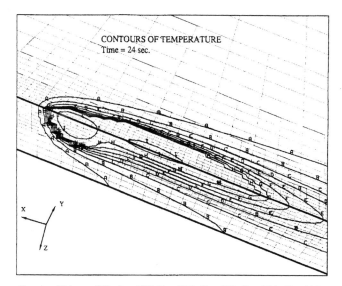

Contour Values (°C): A = 197; B = 375; C = 553; D = 730; E = 908;
F = 1090; G = 1260; H = 1440; I = 1620

Figure 3-10: A closer look at the weld pool and surrounding area. The min. temperature=20.0°C; the max. temperature=1800°C, from [2].

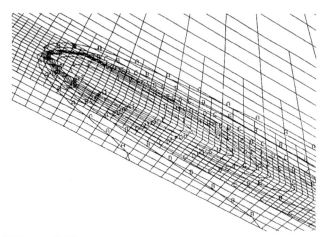

Contour Values (°C):
A=277; B=475; C=673; O=877; E=7070; F=7270; G=7470; H=7600

Figure 3-11: Temperature contours for the corresponding single weld pool model. The min. temperature= 20.0°C; the max. temperature = 2000°C, from [2].

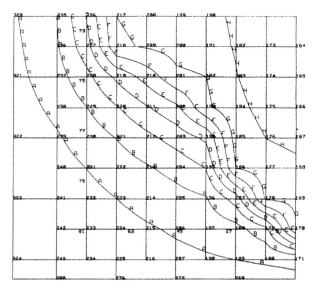

Contour Values (°C): A = 553; B = 709;   C = 865;   D = 1020;
E = 1180; F = 1330; G = 1490; H = 1640

Figure 3-12: The envelope temperature contours on a section on which the temperature field projects. The numbers shown on the mesh are FEM node numbers, from [2].

Glickstein and Friedman [7] used a two-dimensional model to simulate the temperature field in a section along the weld path. Their model gave similar contours outside of the *HAZ* although it did not predict transient temperatures accurately inside the *HAZ*.

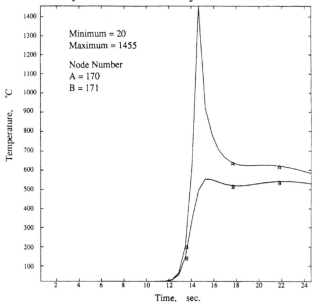

Figure 3-13: The thermal cycles of the nodes directly under the weld path. The node numbers are shown in Figure 3-11, from [2].

Figure 3-12 is the envelope of maximum temperatures in a section, through which the weld pool has passed completely. Note an envelope temperature contour is the locus of maximum temperatures projected on the plane perpendicular to the weld line. It is not the temperature contours on a cross-section at some instant of time. Note the melting point temperature is between contour G and contour F. The contour representing melting temperature envelope is superimposed on the weld pool. There is good agreement with data from experiment.

For nodes 170 and 171, i.e., for the region directly under the weld pool, the thermal cycles are simple. The *HAZ* is thin and basically the same as those obtained from the double ellipsoid heat source simulation. Similarly the thermal cycles of nodes *199, 201,* and *217* at the widest part of the weld pool are not unusual.

Corresponding to the node numbers in Figure 3-12, a set of thermal cycle (time vs. temperature) curves are given in Figures 3-13 to 3-15. The computed thermal cycle within one *FEM* element (nodes *183*, *184*, *192*, and *193*) is obtained by interpolating the curves at corner nodes. It is found that the shape of the thermal cycle is very sensitive to the position selected.

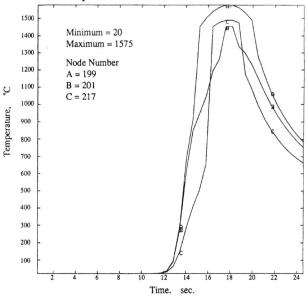

Figure 3-14: The thermal cycles of the nodes along the widest side of the double weld pool. The node numbers are shown in Figure 3-12. These thermal cycles are similar to those obtained from single weld pool, from [2].

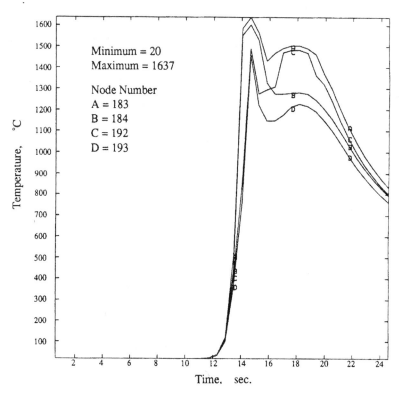

Figure 3-15: The thermal cycles of selected nodes within the bay region. Note the temperature arrest due to the second weld pool in the model, from [2].

The effect of the second weld pool is most apparent in the 'bay area'. The thermal cycles at nodes *183, 184, 192,* and *193* suggest a second temperature peak. Therefore this area is held at high temperature *(> 1100°C)* longer than other areas in the *HAZ*. Because of these quite different thermal cycles, temperature history dependent phenomena are expected to be quite different from other areas of the *HAZ*.

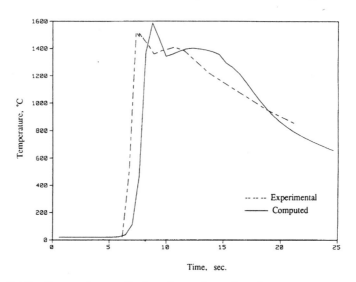

Figure 3-16: Comparison of the thermal cycles between computed and experimental [5] results in bay area. Note that the triggering times are set differently for ease of presentation, from [2].

Figure 3-16 compares experimental [5] and computed results in the bay area. Note that the agreement is excellent, which supports the analysis.

Welds with Filler Metal Addition

Two test cases, of butt welds, have been analyzed on the same mesh structure, shown in Figure 3-17. The characteristics of this mesh are that the liquid metal in the weld pool is excluded and the weld bead can be seen on the plate after the weld pool to account for the added filler material.

In the first test case, the base metal is aluminum which has a high thermal diffusivity and the second is steel which has a low thermal diffusivity. The surface of the weld pool is prescribed at the melting temperatures of the base metals. Both surfaces of the weld joints have been prepared with V grooves. Welds are designed for one pass on each face. The weld pool surface was measured experimentally. First a high pressure gas jet was used to blow the liquid metal out of the weld pool "exactly" when the arc was turned off. Then the weld pool surface was measured directly.

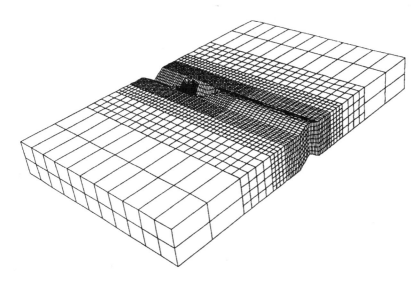

Figure 3-17: The FEM mesh is built on a Eulerian frame. The finest mesh is in the front of the weld pool. The element sizes after the weld pool can be changed in analyses, so that more or less material is included, from [14].

The Eulerian frame is created on a mesh *218mm* long, *165mm* wide and *19mm* thick. The welding speed was *6mm/s*. The length of the weld pool was *32.5mm*. The center of the pool was *10mm* from the leading edge. The maximum pool width was *15mm*. The deepest point was 10.1*mm*. The thickness of the frame is the same as the plate in the real weld, but the length and the width follow that of the sample. The length and the width of the real weld are unknown. Nevertheless, the sample piece cut from the welded plate is assumed to be in steady state. The parameter α was chosen as $c_p/\kappa = 1.76 \times 104\ s/m^2$.

Figure 3-18 shows the computed isothermal contours. Gu et al [14] compares the *FEM* 3-dimensional steady state analysis with Rosenthal's 1-dimensional and 2-dimensional analytical model, Figure 3-19.

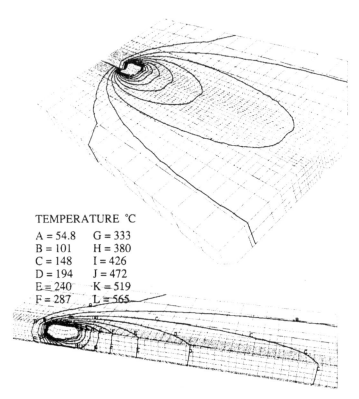

TEMPERATURE °C

A = 54.8	G = 333
B = 101	H = 380
C = 148	I = 426
D = 194	J = 472
E = 240	K = 519
F = 287	L = 565

Figure 3-18: The isothermal contours in the steady state of an arc weld on aluminum plate, computed by a transformed Eulerian formulation. Because of high diffusivity the density of the mesh around the weld pool could be lower than that for steel. (a) Whole mesh; (b) mesh cut along the weld path, from [14].

The second test was performed a low carbon steel plate, Figure 3-20. The weld speed was 1.5 mm/s. Estimated heat input per unit-length was 1.5 *kJ/mm*. The diffusivity of steel is smaller than that of aluminum. Thus, for the same welding speed, the Eulerian domain for steel can be smaller than that for aluminum. However, the mesh around the weld pool should be finer in order to describe the sharper gradient. The main purpose of this analysis was to compare the results of a transformed Eulerian formulation with that of a time marching Lagrangian formulation.

To assess the efficiency, the same problem was analyzed using a Lagrangian formulation, Figure 3-21. Fifty-five time steps were taken to reach a relatively stable steady-state temperature field near the weld pool. The same value of α was used, $\alpha = 10^{5} s/m^{2}$. The meshes for both analyses consisted of about 12000 elements.

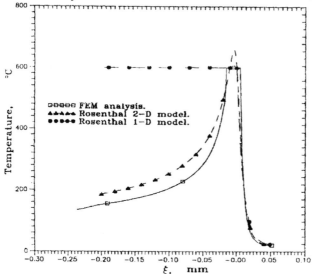

Figure 3-19: Comparison of thermal cycles in steady state of weld on aluminum. Note that the 1-dimensional and 2-dimensional analytical solutions were calculated by Rosenthal's model. Since the 1-dimensional model uses a sectional source and the 2-dimensional model uses a circular through-thickness source, both energy inputs are higher than in reality, from [14].

Comparing Figures 3-20 and 3-21, we can find the difference in the contours on the symmetric plane along the weld path. The contours in Figure 3-21 distribute evenly through the thickness after the weld pool. It is apparent that the filler material plays an important part in this difference.

Since the thermal cycle under the second weld pool is different, the microstructure is expected to be different. From the method of Khoral [23], the thermal analysis is coupled to the microstructure analysis. The results are shown in Figures 3-23 and 3-24.

Since the method is valid only in the *HAZ*, those contours inside the weld pool (contour G, Figure 3-12) have no meaning. Along the weld pool, there is a very thin layer of large grains. As expected, this thin layer becomes much thicker in the bay area. More important the fast cooling rate and high hardenability caused by the large grain size in the bay area tend to cause more bainite and martensite to form, which causes the area to be harder and possibly more brittle.

Figure 3-22 compares the thermal history of points on the top surface parallel to the weld path. Because of the differences in the meshes used in the two formulations, the nodal positions are not identical.

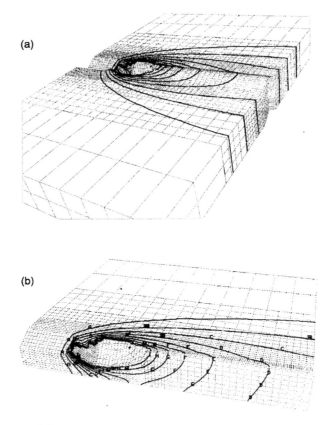

(a)

(b)

Temperature °C
A=144 B=299 C=453 D=607   E=762   F=916 G=1070 H=1230
I=1380

Figure 3-20: The isothermal contours in the steady state of an arc weld on low carbon steel, computed by a transformed Eularian formulation. Because of low diffusivity, the density of the mesh around the weld pool should be higher, especially around the first half of the weld pool. (a) Whole mesh; (b) mesh cut along the weld path, from [14].

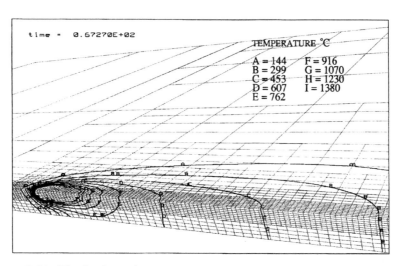

Figure 3-21: The isothermal contours computed by the Lagrangian formulation. The contours represent the solution in 55 time steps, 67 s after striking the weld torch. The velocity and diffusivity used were the same as those used for Figure 3-19. Note that without the addition of filler material the temperature varies less in the thickness direction after the pool passes, from [14].

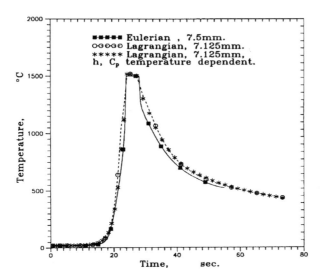

Figure 3-22: Comparison of thermal cycles at a point near the weld pool between the transformed Eulerian formulation and the Lagrangian formulation in steady state. The weld is on low carbon steel. The millimeters indicate the distances from the weld pool.

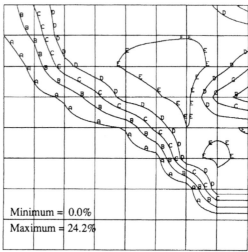

Minimum = 0.0%

Maximum = 24.2%

Contour Values: A = 2.42%; B = 4.83%; C = 7.25%;
D = 9.66%; E = 12.1%; F = 14.5%

Figure 3-23: Bainite contours in a cross-section, where the austenite transformation is complete

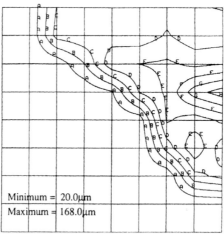

Minimum = 20.0μm

Maximum = 168.0μm

Contour Values (μm): A = 25.8; B = 41.5; C = 57.3;
D = 73.0; E = 88.8; F = 105.0; G = 120.0; H = 136.0

Figure 3-24: Grain size contours in a cross-section, where the temperature has cooled to under 200 °C in the HAZ. The grain size is sensitive to the time the austenite phase spends above the dissolution temperature for NbC and VC precipitates.

A steady state analysis for arc welding with a transformed Eulerian formulation, compared to the usual Lagrangian formulation, has the advantages of low computing cost and high resolution. The model also allows filler material to be added during welding. The test cases indicate that a reasonable accuracy could be achieved with the boundary conditions used. The model can also serve as a good predictor for a full non-linear solver. The thermal cycle of a material point was obtained by traversing geometry space along a flow line and mapping values into the time dimension. The thermal histories obtained can be used to analyze the stress and microstructure in the neighborhood of the weld pool.

Correctness of heat input, i.e., the incoming heat distribution, is critical for the analysis inside the *HAZ*. It can be expected that different shapes of the heat source will result in different temperature contours, microstructures, as well as strain and stress distributions in the *HAZ*.

References

1. Goldak J.A., Patel B., Bibby M. and Moore J Computational weld mechanics, AGARD Conf. Proceedings No. 398, June 1985
2. Gu M., Goldak J.A. and Bibby M.J.; Computational heat transfer in welds with complex weld pool shapes, Adv. Manuf. Eng., Vol. 3. January 1991
3. Lindgren L-E. Finite element modeling and simulation of welding Part I Increased complexity, J of Thermal Stresses 24, pp 141-192, 2001
4. Radaj D. Eigenspannungen und Verzug beim Schweissen, Rechen- und Messverfahren, Fachbuchreihe Schweisstechnik, DVS-Verlag GmbH, Duesseldorf 2000
5. Barlow J.A. One weld or two? The formation of a submergend arc weld pool, Welding Inst. Res. Bull., pp 177-180, June 1982
6. Kraus H.G. Experimental measurement of stationary SS 304, SS 316L and 8630 GTA weld pool surface, Welding J.
7. Glickstein S.S. and Friedman E. Effect of weld pool configuration on heat affected zone shape, Tech. Notes Welding Res. Suppl. Thesis, pp 110-112, June 1981
8. Radaj D. Schweissprozesssimulation, DVS-Verlag, Duesseldorf 1999

9. Jones BK, Emery AF and Marburger J. An analytical and experimental study of the effects of welding parameters in fusion welds, Welding J Vol. 72, No. 2, pp 51s-59s, 1993

10. Jones B.K., Emery A.F. and Marburger J. Design and analysis of a test coupon for fusion welding, ASME J. Pressure Vessel Technology, Vol. 115, pp 38-46, 1993

11. Roelens J.B., Maltrud F. and Lu J. Determination of residual stresses in submerged arc multi-pass welds by means of numerical simulation and comparison with experimental measurements, Welding in the World, Vol. 33, No. 3, pp 152-159, 1994

12. Roelens J.B. Numerical simulation of multipass submerged arc welding-determination of residual stresses and comparison with experimental measurements, Welding in the World, Vol. 35, No. 2, pp 110-117, 1995

13. Lindgren L-E., Runnemalm H. and Naesstroem M.O. Numerical and experimental investigation of multipass welding of a thick plate, Int. J. for Numerical Methods in Engineering, Vol. 44, No. 9, pp 1301-1316, 1999

14. Gu M., Goldak J.A. and Hughes E. Steady state thermal analysis of welds with filler metal addition

15. Goldak J., Chakravarti A. and Bibby M. A finite element model for welding heat sources, Metallurgical Transactions B, Vol. 15B, pp 299-305, June 1984

16. McDill J.M.J. (1988). An adaptive mesh-management algorithm for three-dimensional finite element analysis. PhD thesis, Carleton University.

17. Rykalin R.R. Energy sources for welding, Welding in the World, Vol. 12, No. 9/10, pp 227-248, 1974

18. Rosenthal D. the theory of moving sources of heat and its application to metal treatments, Trans ASME, Vol. 68, pp 849-865, 1946

19. Pavelic V., Tanbakuchi R., Uyehara O. A. and Myers: Experimental and computed temperature histories in gas tungsten arc welding of thin plates, Welding Journal Research Supplement, Vol. 48. pp 295s-305s, 1969

20. Sudnik V.A. and Erofeew W.A. Computerberechnungen von Schweissprozessen, Techn. Hochschule Tula, Schweden, 1986

21. Ohji T., Ohkubo A.and Nishiguchi K. Mathematical modeling of molten pool in arc welding, Mechanical Effects of Welding, pp 207-214, Publ. Springer, Berlin 1992

22. Weiss D., Schmidt J. and Franz U. A model of temperature distribution and weld pool deformation during arc welding, Mathematical Modeling of Weld Phenomena 2, Edited by Cerjak H. and series Editor Bhadeshia H.K.D.H., The Institute of Materials 1995

23. Khoral P (1989). Coupling microstructure to heat transfer computation in weld analysis. Masters Thesis, Carleton University

24. Sudnik W., Radaj D. and Erofeew W. Validation of computerized simulation of welding processes; 4th Int. Seminar on Numerical Analysis of Weldability, Seggau near Graz, 29. Sept. -1 Oct. 1997

25. Sudnik W., Radaj D. and Erofeew W. Computerized simulation of laser beam weld formation comprising joint gaps; J. Phys. D. Appl. Phys. 31, pp 3475-3480, 1998
26. Sudnik W., Radaj D., Breitschwerdt S. and Erofeew W. Numerical simulation of weld pool geometry in laser beam welding; J Phys. D. Appl. Phys. 33, pp 662-671, 2000
27. Radaj D., Sudnik V., Karpuchin E, Berger P. and Heckeler G. Comparison of heat sources for modeling laser beam welding; Publication in Metallurgical and Materials Transactions B

Chapter IV

Evolution of Microstructure Depending on Temperature

4.1 Introduction and Synopsis

One of the most widely used methods for joining metal in engineering is fusion welding. The progress made in the chemical and oil industry, aerospace, shipbuilding, structures, etc. heavily relies on reliable welds. Potential losses and risk to human life involved in weld failure have resulted in stringent material and welding process control requirements.

The energy input causes a thermal cycle that drives the grain growth and microstructural changes in the heat affected zone. The *HAZ* ranges from the solid-liquid transition zone on its inner surface to the unaffected base metal on its outer surface. A large austenite grain size in steel welds along with transformation products such as martensite and bainite renders the base metal susceptible to brittle fracture. In addition these hard and brittle products can make the weld sensitive to stress and corrosion and can also lead to the formation of stress induced cold cracks and pose most danger in practice.

Microstructure computations are also fundamental to predicting the behavior of the *HAZ*. The *HAZ* is the most sensitive area of the weld joint and in steel the mechanical properties of the joint depend on austenite grain size and transformation products.

At a macroscopic level, the metal physics can be described by scalar fields specifying the fraction of each phase present, the composition of each phase and the grain size of each phase at each spatial point in the heat affected zone. In low alloy steels, the phases of greatest interest are ferrite, pearlite, austenite, upper and lower bainite, martensite, carbide and liquid. Most steels used in welded structures are produced with a ferrite-pearlite microstructure or a ferrite and austenite-martensite constituent. When these steels are heated above A_{e1} or the eutectoid temperature, the pearlite or austenite-martensite constituent rapidly transforms to austenite with the same composition. Even though the heating rates are high, superheating only amounts to a few tens of degrees; approximately *80°C* is common. So in most cases the assumption that the transformation occurs at equilibrium can be tolerated. However, the homogenization of this carbon austenite region is much slower. It is a function of the temperature, time and the distance between high carbon regions.

The various regions into which the thermal cycle can be divided are shown in Figure 4-1.

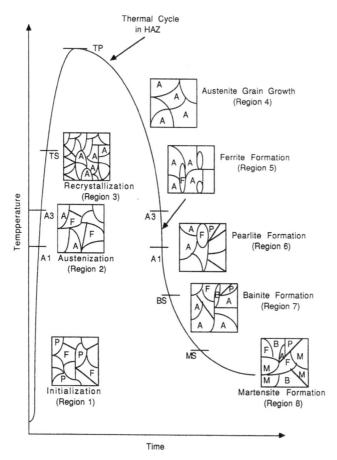

Figure 4-1: Schematic representation of the transformation regions on the thermal cycle, from Khoral [4].

A group of researchers [11, 12, 13 and 14] have attempted to predict the hardness distribution in the *HAZ* without taking into account the phases present in the microstructure. Another approach to the *HAZ* microstructure prediction has been proposed by Ashby, Ion and Esterling [6, 9 and 10].

The weld thermal cycle is predicted by using a modified version of Rosenthal's [15] analytical equation. One drawback of this model is the weakness of the Rosenthal equation, which ignores the distribution of power density and assumes that thermal conductivity and specific heat of the material are independent of temperature.

The thermal conductivity varies as shown in Figure 4-2. Here the conductivity in the liquid has been set at a relatively high value in an attempt to take into account the convective stirring at the high velocities *(~ 1 m/s)* found in weld pools [32].

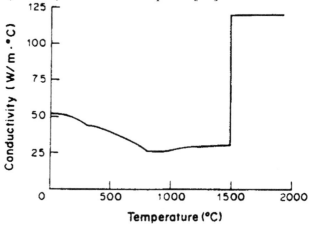

Figure 4-2: The thermal conductivity-temperature relationship for a low alloy steel that was used in the finite element program by Watt et al. [21]. Note that a high conductivity in the molten region is used to simulate heat transfer by stirring.

The heat capacity and the latent heats vary also as shown in Figure 4-3.

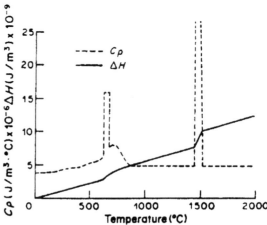

Figure 4-3: The volumetric heat capacity and latent heats of fusion and transformation arranged for finite element computer implementation by Watt et al. [21].

A third set of workers have modified Kirkaldy's hardenability algorithm [16, 17, 18 and 19] to predict the microstructure of the *HAZ*. This algorithm can be used to isolate the individual effects of weld process parameters. It also highlights the effect of austenite grain growth on the *HAZ* microstructure. For example, *CCT* diagrams and Jominy-end-quench hardness curves can be generated from the Kirkaldy system for most low alloy steels. The austenite decomposition relationships in this scheme are formulated as ordinary differential equations *ODE* (see the general form in equation 4-21 and the special form for the austenite transformation to ferrite in equation 2-19) which must be integrated numerically to compute the fraction of each microstructural component formed. This can be *CPU* intensive if a simple Euler integration scheme is used.

An adaptive integration scheme has been suggested to increase computational efficiency [22]. A computed microstructure distribution across a small bead-on–plate weld is shown in Figure 4-4 to demonstrate the sensitivity of the system. Similar computations that are transient are possible because the microstructure is tracked over the entire thermal cycle.

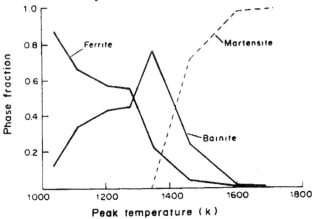

Figure 4-4: Variation of ferrite, bainite and martensite throughout the HAZ of the bead-on-plate weld (base metal composition and operating parameters for low alloy steel weld analyzed: C 0.12, Si 0.16, Mn 0.91, P 0.002, S 0.005, Al 0.04, Nb 0.021, N 0.011, gas metal arc, gross heat input 0.21kJ/mm, efficiency 0.75 and plate thickness 12.7mm)

This group of researchers, [20, 21 and 22], have shown that the hardenability algorithm proposed by Kirkaldy can predict the resultant microstructure of the *HAZ* quite accurately. The main draw back in their computation is the use of inefficient integration techniques to solve the austenite decomposition relationships. This problem limited accuracy and made their algorithm *CPU* intensive and costly.

Although many factors can affect the microstructure of a metal, the thermal cycle is the dominant factor in welding processes. It is the peak temperatures and the heating and cooling rates in the thermal cycle that control the phase changes and microstructures. When the thermal cycle reaches its steady state, the microstructural evolution will, in most cases, also reach a steady state.

4.2 Microstructure Model

The microstructure algorithm used here was motivated by the work of Kirkaldy [16], and originally developed for welding by Watt et al.[21], Henwood et al.[22] and enhanced by Khoral [4].

The various microstructural transformation temperatures are a function of carbon and alloy content of the low alloy steel.

For low alloy steel, the liquidus and solidus temperature can be determined as a function of carbon content [35].

$$T_L = 1530.0 - 80.581C \qquad (4-1)$$
$$T_S = 1527.0 - 181.356C \qquad (4-2)$$

where

C $\quad$ = carbon content of the steel

T_L $\quad$ = liquidus temperature of the steel (°C)

T_s $\quad$ = solidus temperature of the steel (°C)

In modern micro-alloyed steels, austenite grains are pinned by carbide/nitride precipitates. Ashby, Easterling and Ion [6, 9 and 10] presented a modified relationship to calculate the precipitate dissolution temperature. This relationship takes into account the effect of these grain growth retarding precipitates and is given as:

$$TS = \frac{B}{A - \log[C_m{}^a C_c{}^b]} \tag{4-3}$$

where

$A\&B$ = constants that depend on the precipitate species. $A=3.11$ and $B=7520.0$ from [4 and 36]

C_m & C_c = concentration of metal (*Nb, Ti, V* etc.) and non metal (carbide, nitride or boride) in precipitate respectively

a & b = stoichiometry constants. $a=1,0$ and $b=1,0$ for *VC* from [4 and 36]

The concentrations are in wt % and temperature is in degree Kelvin.

The upper critical temperature A_3 depends on the carbon content of the steel as shown in the iron-carbon phase diagram Figure 2-30. The most accurate representation for the A_3 line is given by Kirkaldy and Baganis [19]. But this is not convenient in the computer model because of the difficulty involved in inverting this relation for use in the lever law. To avoid this difficulty an alternative relation proposed by Leslie [33] is used in this algorithm.

$$A_3(°C)=912-200\sqrt{C}-15.2Ni+44.7Si+315Mo$$
$$+13.1W-(30Mn+11Cr+20Cu-700P-400Al-120As-400Ti) \tag{4-4}$$

where the composition are in wt %.

The lower critical temperature (A_1) or pearlite start temperature is also dependent on the composition of the low alloy steel and is given as [34]:

$$A_1(°C)=723-10.7Mn-16.9Ni+29Si+16.9Cr+290As+6.4W \tag{4-5}$$

As the weld cools and the temperature drops below the bainite start temperature *(BS)*, the formation of pearlite and ferrite stops and the austenite decomposition to bainite starts. The bainite start temperature is given as [16]:

$$BS(°C) = 656 - 58C - 35Mn - 75Si - 15Ni - 34Cr - 41Mo \tag{4-6}$$

On rapid cooling of the weld, austenite in the heat affected zone starts decomposition into martensite. This transformation starts below the martensite start temperature, which is given as [34]:

$$MS(°C) = 561 - 474C - 35Mn - 17Ni - 17Cr - 21Mo \tag{4-7}$$

The carbon content of eutectoid alloy is also a constant for a particular low alloy steel and is given by the following relations [33]:

$$C_{eut} = \frac{[\phi_1 - \phi_2 - A_1]^2}{203^2} \tag{4-8}$$

$$\phi_1 = 910 - 15.2Ni + 44.7Si + 104V + 315Mo + 13.1W \tag{4-9}$$

$$\phi_2 = 30Mn + 11Cr + 20Cu - 700P - 400Al - 120As - 400Ti \tag{4-10}$$

All these parameters given by equations (4-1) to (4-10) are used to divide the *HAZ* thermal cycle into eight distinct regions as shown in Figure 2-30.

Phase transformations taking place in the *HAZ* during the weld thermal cycle can be divided into two sections a) transformations that take place during heating and b) transformations that take place during cooling.

Initialization of ferrite and pearlite spans from room or pre-heat temperature to the lower critical temperature A_1. In this region, in the absence of knowledge of the real microstructure, the microstructure is assumed to be in equilibrium and is a mixture of ferrite and pearlite. The volume fraction of ferrite and pearlite is given by the lever law [4]:

$$X_F = \frac{C - C_{eut}}{C_\alpha - C_{eut}} \tag{4-11}$$

$$X_P = 1 - X_F \tag{4-12}$$

where

X_F = fraction of ferrite

X_P = fraction of pearlite

C   = carbon content of steel

C_{eut} = carbon content of the eutectoid

C_α = carbon content of ferrite

The carbon content of ferrite below A_1 is given by an empirical equation assuming a linear decrease from eutectoid value to a value of zero at room temperature *(20 °C)* [4 and 20]:

$$C_\alpha = \frac{T - 20.0}{A_1 - 20.0}(0.105 - 115.3 \times 10^{-6} \times A_1) \tag{4-13}$$

As the temperature in the *HAZ* exceeds the A_1 line during heating, pearlite colonies transform to austenite. The intercritical region or partially transformed region of the *HAZ* is the region where temperature varies between A_1 and A_3. As the peak temperature goes towards A_3, the austenite content increases under near equilibrium conditions. Because of the very high heating rate in the welds α-phase is not likely to get superheated substantially before the α → γ phase transformation takes place. Under equilibrium conditions the amount of austenite and ferrite is given as:

$$X_F = \frac{C - C\gamma}{C_\alpha - C_\gamma} \tag{4-14}$$

$$X_A = 1 - X_F \tag{4-15}$$

where

X_A = fraction of austenite

C_γ = carbon content of austenite

C_α = carbon content of ferrite

The carbon content of austenite can be obtained by rearranging equation (4-4):

$$C_\gamma = \frac{[\phi_1 - \phi_2 - T]^2}{203^2} \tag{4-16}$$

$$\phi_1 = 910 - 15.2Ni + 44.7Si + 104V + 315Mo + 13.1W \tag{4-17}$$

$$\phi_2 = 30Mn + 11Cr + 20Cu - 700P - 400Al - 120As - 400Ti \tag{4-18}$$

Above A_1 the carbon content of ferrite is given by the following relation:

$$C_\alpha = 0.105 - 115.0 \times 10^{-6} \times T \tag{4-19}$$

As the peak temperature goes above A_3 the remaining ferrite matrix also transforms to austenite. Thermal cycles with peak temperature between A_3 and precipitate dissolution temperature represent the recrystalized zone.

The growth in the grain size of the austenite that forms can be described as a function of temperature and time. However, growth will not begin until carbon-nitride precipitates such as *VC* and *NbC* (niobium) dissolve.

It is assumed that grain growth is diffusion controlled and the driving force is the surface energy and it does not require any nucleation. The grain growth equation is given as [10]:

$$\frac{dg}{dt} = \frac{k}{2g}\exp(-\frac{Q}{RT})$$ (4-20)

where

g = grain size *(μm)*
k = grain growth constant *(μm²/s)*
Q = activation energy for grain growth *(kJ/mol)*
R = universal gas constant *(kJ/mol K)*
T = temperature *(°K)*
t = time *(s)*

Q and k are dependent on the type of precipitate [4 and 10]. The austenite grain growth which begins at precipitate dissolution continues up to the peak temperature of the thermal cycle.

On cooling as the temperature drops below the A_3 line, austenite starts decomposing into its daughter products. The kinetics of austenite decomposition into its daughter products are described by ordinary differential equations *(ODE)*, [4, 21 and 22], of the form:

$$\frac{dX}{dt} = B(G,T)\,X^m(1-X)^p$$ 4-21)

where

X = volume fraction of the daughter product
B = effective rate coefficient
G = austenite grain size
m,p = semi-empirical coefficients, were determined by experiment by Kirkaldy [16].

The term $B(G,T)$ is a function of grain size, under-cooling, the dependence of the carbon diffusivity on the alloy and temperature and the phase fractions present.

This is essentially the model presented by Watt et al [21]. The microstructure algorithm has been coupled with the three dimensional finite element heat transfer program developed by Goldak et al [26 and 27] to predict the transient microstructure of the *HAZ*.

The mechanical properties of metals are sensitive to their microstructure. By intentionally changing the microstructure, one can control the properties to provide the best service. However, not all the changes are beneficial. Many manufacturing processes can produce unfavorable microstructures, which can reduce the reliability and performance of the product. In welding, undesirable microstructures can cause failures and are a serious concern. Whether the microstructural changes bring fortune or catastrophe will depend on one's capability to predict why and how these changes happen. The prediction of the microstructural changes caused by thermo-mechanical processing continues to be an area of great practical benefit.

The major contribution of this chapter to the simulation of microstructures is to create a data structure for the coupling of a steady state thermal field to the formulation of microstructures, Gu et al. [23]. Some changes have been made in order to satisfy conditions in the weld pool and fusion region.

Methods to predict the steady state temperature field around a weld pool were first reported by Rosenthal [7] and Rykalin [8].

In the previous chapter, a *FEM* model of a steady state temperature field around the weld pool was described. Since this is a true three-dimensional model, it can investigate different welds in detail for their unique temperature distributions. Such distributions are sometimes critical to the simulation of microstructures. Compared to a time marching Lagrangian formulation for a transient thermal analysis, the cost of computing a steady state temperature field has been reduced from hours or days to seconds. The difficulty of coupling the steady state temperature field with a microstructure prediction is to extract the thermal cycle from the steady state temperature field, especially when the mesh is complex, i.e. unstructured.

Microstructure evolution is associated with a material point. In a Lagrangian formulation, nodes and Gauss points are material points. In this case, one simply solves the equations for microstructure evolution as a function of temperature, time and possibly stress for any or all nodal or Gauss points. However, in an Eulerian formulation, e.g., a steady state formulation for a moving weld, the nodal and Gauss points are spatial points and not material points. In this case, one must first compute the curves traced out in space for each material point for which one wishes to compute the microstructure evolution. In the general case, where the velocity field is a function of time, such curves are called flow lines. At each instant of time, the tangent of the curve is parallel to the velocity field. In the steady state case, the velocity field is not a function of time. In this case, the tangent of each point on a curve is parallel to the constant velocity field. Such curves are called streamlines.

In an arbitrary Eulerian Lagrangian (*AEL*) formulation, the mesh is allowed to move with the material points or with spatial points or allowed any other convenient mesh motion or velocity. In this case, the task of computing flow lines or streamlines is not fundamentally different from the Eulerian case. One simply deals with both the velocity of material points and the velocity of the mesh.

To capture the thermal cycle in a steady state temperature field, the applied data structure should meet several requirements. The data structure must support efficient ordered traversals of the spatial points on a streamline (starting at the head and proceeding downstream). It must also allow the element that contains the head of a streamline to be found. Lohner used a heap of faces [24], supporting insertion and deletion in $O(\log_2 n)$ operations for n faces. However, given different sized faces this may require a complete search to find the face containing a point. A data structure has been developed by Gu et al. [25] to represent properly the streamlines in a steady state field and to meet all the above requirements.

In a steady state temperature field described in a Eulerian frame, flow lines and streamlines are equivalent. Integration along a flow line always gives the thermal history of a material point. In a general *FEM* mesh with refinement around the weld pool, the creation of flow lines can consume more time than the microstructure calculation itself. Data structures to organize elements, points and flow lines are necessary. The functionality of the data structure is critical to the success of the model.

4.2.1 Data Structures

Flow lines

The computation of microstructures at a spatial point requires that the thermal history be deduced from a set of elements or nodes in a steady state temperature field. The correct history of a material point at a downstream spatial point can be revealed only by following a flow line from its upstream origin to its downstream destination. For this reason, the flow line collection is organized as described below, and as shown in Figure 4-5.

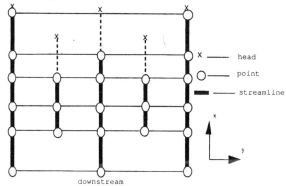

Figure 4-5: The structure of flow lines in a two-dimensional cross-section of a three-dimensional FEM mesh.

(1) Each flow line has three parts: head, body, and front end condition. The head is initial position. Front end condition is initial condition. The body is the flow line path described by a collection of points, ordered from upstream to downstream. Every point Xi with coordinates (x_i, y_i, z_i) in the body stores a pointer to the point, $X_{i-1}(x_{i-1}, y_{i-1}, z_{i-1})$, immediately preceding it. The time increment Δt is easily calculated, since the velocity v is known. In this context, v is the weld speed, the weld path is assumed to be along the x-axis, and v and x_i are scalars.

$$\Delta t = \frac{\Delta x}{v} = \frac{\Delta x_i - \Delta x_{i-1}}{v} \qquad (4\text{-}22)$$

The front end condition can be either defined from the problem data, if the flow line starts at a surface of the domain, or interpolated from the computed data within the element in which the flow line starts.

(2) The flow lines are kept in a global collection, sorted by the x-coordinate of their head positions. The computation loops over the flow lines in the sorted order, proceeding downstream along each flow line. This structural order guarantees the computing order that is necessary when a given point "p" represents the start condition of a flow line. Only after the values of all the points, contained in an upstream element of the point "p" have been computed can a value at "p" be computed.

To satisfy the requirements in building the above collection of flow lines, Gu et al [2] created a multi-layered tree structure as described later.

Functionality of the Data Structure

Some functionality will be required in the process to build up the flow lines. Since there are some similarities among all requirements, the required functionality will be implemented in one data structure. The data structure provides the functionality to:

1. loop over elements in the order from upstream to downstream;
2. find an element containing a given point;
3. find all points from a given set lying along a given flow line.

Smalltalk has been used in the implementation of the functionality. With the help of Object Oriented Programming, the data structure has not only been created easily, but organized nicely as well.

Layered tree description

The structure chosen can be viewed as a three-layered tree, used to hold either points on a flow line or elements. The layers do not refer to the levels of a regular binary tree. Rather, each layer consists of one or more complete trees, each one hanging off a vertex of a higher level tree. The layers represent the x-, y-, and z-axes, respectively, and the ordering within a layer is with respect to the corresponding axis.

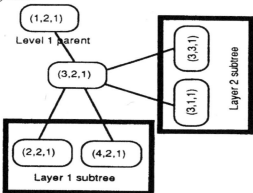

Figure 4-6: An example of the tree structure in two dimensions. The vertex (3, 2,1) is part of a layer 1 search tree based on x-coordinates, but is also the root of a layer 2 subtree based on y-coordinates.

A concrete example with two layers is shown in Figure 4-6. On the left is the top-level tree, a binary search tree based on x-coordinates. The vertex (3,2,1) lies somewhere in the middle of this tree, with parent (1,2, 1) and children (2,2,1) and (4,2,1). However, (3,2,1) is also the root of another tree, which is in the second layer, and whose search ordering is based on y-coordinates. This layer 2 tree contains (in addition to the root (3,2,1)) the points (3,1,1) and (3,3,1). There may be many such layer 2 trees, each one independent. Furthermore, each vertex of a layer 2 tree may have a subtree in layer 3, whose search order is based on z-coordinates.

The structure may also be viewed as a tree of trees. It has the following useful property when organizing points into flow lines. Every vertex of a tree in layer 2 is the root of a layer 3 tree containing points with identical x- and y-coordinates sorted in order by a z-coordinate. This is precisely the body of a flow line in the z-direction. Similarly, by skipping the first layer, or by changing the layer ordering, we can obtain the points on a flow line in the x-direction.

Searching a point in a binary search tree costs $O(\log_2 n)$ operations for n nodes or n equally sized elements. When element sizes vary in a mesh, a binary search tree may need an exhaustive search for an element containing a particular point. In this case, a binary tree may be less efficient than a simple sorted list. While some form of balanced binary search trees would probably provide the best performance in searching a node, quite good results in searching both nodes and elements have been achieved in the current implementation using only sorted lists. Figure 4-7 shows the current implementation of the structure in all three levels, with the different layers represented as sorted lists.

The structure in Figure 4-7 can be used to organize elements in the order that they occur along flow lines. For this purpose we make the simplifying assumption that elements can be replaced for most calculations by their axis-parallel bounding boxes. For many elements in these problems this provides a good approximation of their boundary. For the remainder, it can still reduce the number of elements for which the more expensive calculations are required, and help postpone such calculations as long as possible. This is a common technique in geometric algorithms [28].

The tree ordering in this case is based on the maximum x-, y-, or z-coordinate of the bounding box, depending upon which layer of the tree is being processed. In addition, each vertex of the tree stores the maximum extent (in the appropriate direction) of any of the elements in its subtrees. This allows us to consider only the relevant subtrees in most cases. Note that if the weld source moves in a positive x-direction, the flow lines flow in the negative x-direction, the terms "maximum x" and "upstream" will be synonymous.

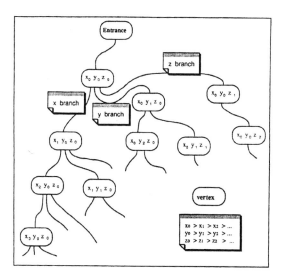

Figure 4-7: Three-layered data structure developed to store elements and to represent flow lines. When an element is stored X, Y and Z are the maximum positions of the element's bounding box. For flow lines X, Y and Z are the coordinates of the head.

Locating elements

The most time-consuming part of the flow line construction is in locating the elements that contain points on the flow line, since there may be several such searches required for each flow line. These searches are required to establish the starting conditions.

At each layer, Gu et al [2] isolate the intervals in the direction for that layer that might contain the query point. For example, in the first (x) layer, with the query point $q = (q_x, q_y, q_z)$ and with vertex V_i, let $x_{\min i}$ be the minimum x-coordinate of an element stored in the vertex V_i and let x_{length_i} be the largest length in the x-direction of all elements in the branches at V_i. Then, in searching for the element containing q, we search only those vertices and branches with $x_{\min i} \leq q_x$ and $x_{\min i} + x_{length_i} \geq q_x$. This process is then repeated for the y- and z-coordinates in the lower layers of the structure.

Finally, the possible elements containing a point (X_0, Y_0, Z_0) will be reduced to a small number for which precise inclusion tests must be done. This is done by solving the nonlinear equations for *r*, *s*, and *t*.

$$g_x = X_0 - N_i(r,s,t)X_i = 0$$
$$g_y = Y_0 - N_i(r,s,t)Y_i = 0 \qquad\qquad (4\text{-}23)$$
$$g_z = Z_0 - N_i(r,s,t)Z_i = 0$$

Using Newton's method [29]:

$$J^k(\xi^{k+1} - \xi^k) = -g^k \qquad\qquad (4\text{-}24)$$

where *J* is the Jacobian matrix, ξ is the vector *(r, s, t)* in the curvilinear coordinate system, and *g* is functions $g = (g_x, g_y, g_z)$. If we write $dr = r^{k+1} - r^k$, $dX = g_x$ and so on for *s, t, y* and *z*, Newton's method can be written in the form convenient in *FEM*:

$$\begin{bmatrix} dx \\ dy \\ dz \end{bmatrix} = J^k \begin{bmatrix} dr \\ ds \\ dt \end{bmatrix} \qquad\qquad (4\text{-}25)$$

To ensure the convergence and improve the solution speed, we project the point first onto the coordinate planes and check whether it is inside the projections of the element bounding box.

Creating flow lines

Flow lines are defined by a set of points. The points can be element nodes or Gauss quadrature points inside elements. Because the velocity is chosen parallel to the *x*-axis in this example, the flow lines are always parallel to the *x*-axis.

The first step in constructing flow lines loops over all points. The layered tree structure is used to organize the points into flow lines as described above. Having the *x*-layer of the structure free, the structure fills its vertices with the points according to their y- and z-coordinates. After all points have been installed, each vertex contains a flow line with one or more points.

A controller or tester is set to prevent a flow line passing through an element without including the pointer to the element. This can be the case when all the element edges are not parallel to the coordinate axes. If this happened, the history of the flow line would be not only incomplete but also inconsistent. The distance between a preceding point and current point is calculated and compared to the length of the element. The distance should not exceed the length.

After flow lines are created, they are collected and sorted according to the position of the head of each flow line, as shown in Figure 4-8. The starting conditions of each flow line can then be found as described below.

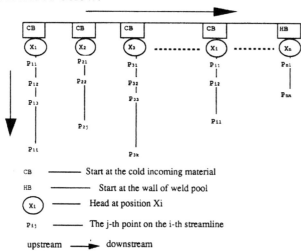

CB	————	Start at the cold incoming material
HB	————	Start at the wall of weld pool
X_i	————	Head at position Xi
P_{ij}	————	The j-th point on the i-th streamline
upstream	——→	downstream

Figure 4-8: A collection of flow lines. The heads are arranged such that $X_1 \geq X_2 \geq X_3 \geq \cdots \geq X_i \geq \cdots \geq X_n$. The number of points in each flow line need not be equal. The arrows indicate the downstream

By using the technique described above, find an element containing the head of the flow line. Take the maximum x-coordinate of this element plus an increment as a new point, and find a new element containing this new point. The new element is the upstream element of the former one and the new point is the predecessor of the head of the flow line in the direction of flow. In the order in Figure 4-8 all points of the new element would have been calculated before the new flow line starts. The new element can thus be used as an interpolation basis for the starting condition.

If no such element is found, then the flow line must start in an element on the free surface of the mesh, and a free surface condition can be applied. If the surface is an upstream boundary then cold material with the initial microstructure can be assumed. If the surface is the wall of a weld pool, then a hot material with a pure austenite or delta ferrite structure can be assumed.

Figure 4-9 shows what flow lines look like in a real discriticized domain.

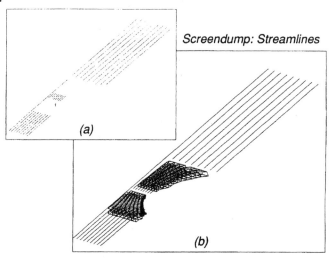

Figure 4-9: Numerical representation of all flow lines (streamlines) on a y = 0 plane. The lines are the links of points in the x-direction. (a) Only the bodies of flow lines are shown. The unstructured mesh with refinement around the weld pool results in short lines. A controller prevents flow lines jumping over any element. (b) Those flow lines starting inside an element get their initial data by interpolating from upstream elements which are stored in heads.

4.2.2 Test Problems and Results

The microstructures were computed first for the temperature field, Figure 4-10, obtained from ref. [25].

The steady state temperature field was computed on a mesh with about 12,000 elements. The welding speed was *1.5 mm/s*. Weld pool size and shape were measured from an experimental weld. The maximum length was *35 mm*, the maximum width *22 mm*, and the maximum depth *11 mm*. The gross power input in this weld would be approximately 1.5 *kJ/mm*. The test plate size was *150 x 40 x 19 mm* and the composition is given in Table 4-1.

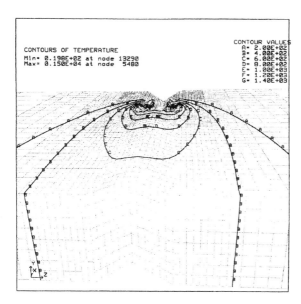

Figure 4-10: Temperature field in a weld computed by a steady state thermal analysis [25]. The microstructures in Figures 4-11,-12 and -14 were computed for this field.

Table 4-1: Chemical composition of the low alloy steel analyzed

C	Si	Mn	P	S	Al	Nb	N
0.12	0.16	0.91	0.002	0.005	0.04	0.021	0.011

The second modeling was performed on a temperature field, Figure 4-15 taken from the transient thermal analysis [12]. A double weld pool phenomenon in this weld was described by Harlow [13]. Special treatment of the boundary conditions, as presented in ref. [12], was used. The speed for this weld was also *1.5 mm/s*. The test plate size was *250 x 160 x 19 mm*. The composition was the same as the first test case.

Before welding, the initial microstructures for both welds were 20% pearlite and 80% ferrite. The initial grain size of austenite was assumed to be 10 *μm*.

Steady State Temperature Field

The Figures in this section show the state at the rear end of the mesh or flow lines. In the regular mesh next to the filler material, each element has a depth of *2.375 mm* in the *y-* direction.

In the upstream area, there is no change from the initial microstructures, since the incoming material has not been affected by the heat input from this weld pool. For this temperature field, in the downstream area, all the austenite has completely decomposed into its transformation products. It could not be seen at the downstream end. This full decomposition indicates that the microstructure of the material flowing out is essentially stable. Although some of these transformation products are metastable, any changes which occur only do so very slowly unless there is further heating.

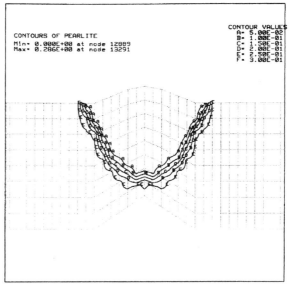

Figure 4-11: Pearlite contours at the ends of streamlines. Contour B is typical of the edge of the HAZ

Figures 4-11 and -12 show the isopercentage contours of pearlite and bainite distributions, respectively. After welding, the fraction of ferrite is almost constant at 70%. The major microstructural change is that the formation of bainite reduces the pearlite percentage in the

HAZ. In this weld, the cooling rate behind the weld pool is not large enough to form martensite at the downstream end. The final microstructure outside the *HAZ* changes little. Inside the *HAZ* it is mostly ferrite with a mixture of pearlite and bainite. There is about *20-25%* bainite adjacent or near the fusion zone, largely due to the increased austenite grain size. Two to three mm within contour E in Figure 4-12, a layer with the largest grain size and the highest bainite is predicted.

The cooling rate is almost constant for an isothermal contour.

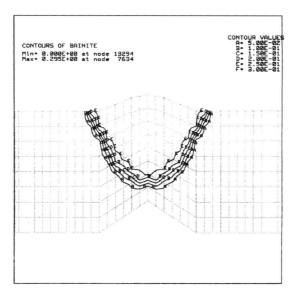

Figure 4-12: Bainite contours at the ends of flow lines. The amount of bainite in the HAZ within contour E is more than 25%.

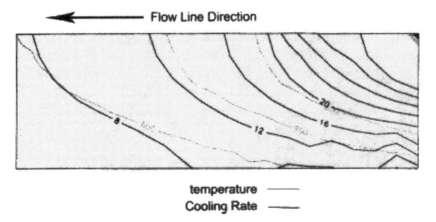

temperature ————
Cooling Rate ————

Figure 4-13: Overlapping cooling rate and temperature contours (grayed). The range shown is 800-600°C. The corresponding range of cooling rate is 20-8°C/s. Note that for an isothermal contour, e.g. 600°C, the cooling rate at that temperature is nearly constant.

The cooling rate is almost constant for an isothermal contour as observed in Figure 4-13.

The cooling rate defined here is *ΔT(°C)/Δt(s)* between two neighboring points along a streamline.

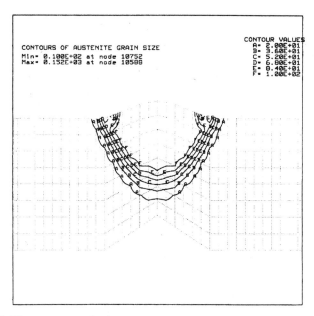

CONTOURS OF AUSTENITE GRAIN SIZE
Min= 0.100E+02 at node 10752
Max= 0.152E+03 at node 10588

CONTOUR VALUES
A= 2.00E+01
B= 3.60E+01
C= 5.20E+01
D= 6.80E+01
E= 8.40E+01
F= 1.00E+02

Figure 4-14: The contours of prior austenite grain sizes (μm) are shown. The grain size in the fusion zone is not shown, since the microstructure model does not apply to the fusion zone. Contours B, C, D, E, and F represent the grain sizes in the HAZ, which are 4-10 times coarser than that in the original base metal.

Figure 4-14 shows the estimated austenite grain size (*μm*). The coarse grained *HAZ* region is next to the fusion zone. The grains in the HAZ are up to *10* times coarser than the original material. Note that the high values inside the fusion zone, or weld metal, may not be correct because this microstructure model does not have the ability to predict them. The *HAZ* region is likely brittle since the upper bainite is expected to form with large grain size at low cooling rate. If lower bainite is formed at higher cooling rates, the region can have an ideal structure, being hard and tough. However, if the cooling rate was high enough to form martensite the region would likely be very hard, brittle and susceptible to cracking.

Clipped transient temperature field

Sometimes an estimate of the steady state microstructures is desired in a temperature field computed by a transient time marching analysis. This kind of temperature field is frequently available in published data. To estimate the microstructures is to explore further the values of the old calculations. This test was performed on a mesh clipped from the example in ref. [30].

About *24* seconds after starting the weld, the weld pool and its surrounding temperature field became stable. The mesh with only the steady state temperature field was extracted, reducing the mesh size by two-thirds.

In a transient study, microstructures are computed after the transient temperature analysis from step to step. After a certain number of time steps the temperature field approaches the steady state in a finite domain near the weld pool. In this domain, however, the computed microstructures may not have necessarily reached a steady state. Before the temperature reaches steady state, the cooling rates are usually higher. The austenite grains tend to grow less. Transformation products that are sensitive to high cooling rates are more likely to form. For these reasons, a Lagrangian formulation will have to run longer to reach a steady state for microstructures than to reach a steady state for the temperature field. Because microstructures are more sensitive to the history, the steady state microstructure forms later than the steady state temperature field. This delay in the time dimension corresponds to a length in the selected spatial domain.

TEMPERATURE °C

Min. = 20	Max. = 1800
A = 300	B = 500
C = 700	D = 900
E = 1100	F = 1300
G = 1500	

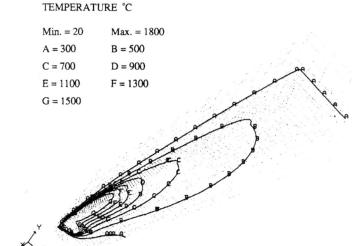

Figure 4-15: The temperature field in a weld with a complex weld pool is shown. It has been computed in ref. [30]. The mesh with these steady state contours was cut from a larger mesh in a transient time marching analysis. The microstructures in Figure 4-16 were computed for this temperature field.

The temperature field adopted is shown in Figure 4-15. Temperature contours in this domain are stable after *62* steps in a Lagrangian calculation. An important characteristic of this temperature field is a "bay region", as described in refs [30] and [31]. In the bay area, material stays above the precipitate dissolution temperature, *1044°C* for this material, for a longer period of time and cools later but possibly faster after the weld pool has passed.

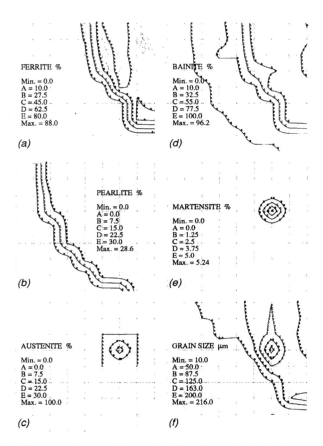

Figure 4-16. Microstructures at the end of flow lines. Only the interesting part at the end of the mesh (Figure 4-14) is shown. A "bay area" due to the complex weld pool shape results in an unusual microstructure. The "bay area" can be seen clearly in (c), (e), and (f), where martensite nucleates, residual austenite remains and the grain size is larger.

Figure 4-16 shows the microstructures at the downstream end, where material has experienced thermal cycles. At the end, the change of pearlite forms an outer boundary to separate the unaffected base metal from the *HAZ*, Figure 4-16b. There is no change to the left of contour E. To the right of contour B, a small amount of pearlite exists. In the *HAZ* right to the contour B, especially in the bay area, transformation products are bainite, Figure 4-16d, as well as some martensite, Figure 4-16e. The modeling of steady state microstructures here predicts that the bay is a hard zone with the potential to initiate fractures. In the bay area, the austenite grain size, Figure 4-16f, is large, and so is the percentage of bainite and martensite. It is the only place that martensite nucleates under the current weld process conditions. Figure 4-16c shows that the decomposition of austenite has not yet reached completion in the bay area because of the truncation of the temperature field. For the existing conditions, the approximately *15%* remaining austenite is likely to transform into martensite at lower temperatures.

The modeling of microstructures in a steady state weld analysis has been implemented. Relative to the Lagrangian formulations, this formulation reduces computing costs and enables higher spatial resolution to be obtained. The model also allows filler material to be added in a natural manner during welding. The test cases indicate that a reasonable accuracy could be achieved with the boundary conditions used. The model can also serve as a good predictor of a full non-linear solver. Using the current model to compute the evolution of microstructures in the steady state of a long weld reduces computing steps from hundred to one and savings in computing costs are significant. The results of the steady state analysis are also believed to be more reliable.

A data structure has been created to support this modeling. The data structure provides flow lines, which in turn provide the temperature history for every point analyzed. The data structure can also be used to find an element that determines the starting condition of a flow line. The functionality of the data structure is critical to the success of the model.

Streamlines play an important role. In a steady state temperature field described in a Eularian frame, integration along a streamline gives the thermal history of a material point. In a general *FEM* mesh with refinement around the weld pool, the creation of streamlines consumes more time than the microstructure calculation itself.

Since the steady state exists only in welds of simple geometry, e.g., prisms or spiral welds on pipes, the work done here focuses only on welding plates. However, the code developed can be used for prisms with any cross-sectional shape. It can also be used when the plate bends into a circular arc, as in the case of welding large diameter pipes. The only change necessary is to use a polar coordinate system for the streamlines. Note that this change may not be applicable for thermal analysis.

4.3 Hardness Calculation of the HAZ

The hardness of the *HAZ* is a very good indicator of its susceptibility to cracks and other problems. The hardness at any point in the *HAZ* can be calculated using the rule of mixtures. Once the volume fractions of ferrite, pearlite, bainite, martensite and austenite are known, then the rule of mixtures can be applied as follows to calculate the hardness [9]:

$$H = H_M X_M + H_B X_B + H_{AFP} X_{AFP} \qquad (4\text{-}26)$$

where

H Total hardness in *VPN*

H_M Vickers hardness of martensite

H_B Vickers hardness of bainite

H_{AFP} Vickers hardness of austenite-ferrite-pearlite mixture

X_M, X_B and X_{AFP} are the volume fraction of martensite, bainite and ferrite-pearlite-austenite mixture respectively. The relationship for hardness calculations of different phases are taken from the work of Maynier et al. [37] and Khoral [4]:

$$H_M = 127 + 949C + 27Si + 11Mn + 8Ni + 16Cr + 21(\log V_1) \qquad (4\text{-}27)$$

$$H_B = -323 + 185C + 330Si + 153Mn + 65Ni + 144Cr + 191Mo$$
$$+ [89 + 53C - 55Si - 22Mn - 10Ni - 20Cr - 33Mo] \log V_1 \qquad (4\text{-}28)$$

$$H_{AFP} = 42 - 223C + 53Si + 30Mn + 12.6Ni + 7Cr + 19Mo$$
$$+ [10 - 19Si + 4Ni + 8Cr + 130V] \log V_1 \qquad (4\text{-}29)$$

All compositions are in wt%. V_1 is the cooling rate at *700°C* *(°C/h)* and can be obtained from cooling time from *800* to *500°C* ($\tau_{8/5}$):

$$V_1 = \left(\frac{800 - 500}{\tau_{8/5}} \right) \left(\frac{1}{3600} \right) \qquad (4\text{-}30)$$

The value of $\tau_{8/5}$ can be measured experimentally or computed from *FEM* or calculated from the Adams [38] relationship for thick plate as follows:

$$\tau_{8/5} = \frac{q/v}{2\pi\kappa}\left(\frac{1}{500-T_0} - \frac{1}{800-T_0}\right) \tag{4-31}$$

where

T_0 The ambient or preheat temperature (°C)

κ The thermal conductivity (J/ms °C)

q/v The heat input (J/m)

It is important to recognize the limits of these equations: 0.1wt%< C < 0.5wt%, Si < 1wt%, Mn < 2wt%, Ni < 4wt%, Mo < 1wt% Cr < 3wt%, V < 0.2wt%, Cu 0.5wt%, Mn+Ni+Mo < 5wt% and 0.01wt% < Al < 0.05wt%. They can only be applied to low carbon steels with reservations, and the micro-alloy elements Nb,Ti,Zr and B are not taken into account.

Good agreement between computed and measured hardness were obtained as shown in results by Henwood et al [22]. There is a relative sharp change in the hardness distribution where the peak temperature is approximately *1400°K*. Since hardness is very sensitive to the martensite fraction, this would imply that the martensite fraction changes rapidly at this same position in the *HAZ* contrary to the measurements. A rapid change in martensite implies a considerable change in hardenability. The austenite grain growth causes martensite not vice versa.

References

1. Goldak J.A. and Gu M. Computational weld mechanics of the steady state, Mathematical Modeling of Weld Phenomena 2, Ed. H. Cerjak, The Institute of Metals, pp 207-225, 1995
2. Gu M.,Goldak J.A. Steady state thermal analysis of welds with filler metal addition, Can. Met., Vol. 32, pp 49-55,1993
3. Rappaz M. Process modeling and microstructure, Philosophical Transactions of the Royal Society of London Series A-Physical Sciences and Engineering, Vol. 351, No. 1697, pp 563 -577, June 15 1995

4. Khoral P. (1989). Coupling microstructure to heat transfer computation in weld analysis. Masters Thesis, Carleton University.
5. Berkhout C.F. Weld thermal simulators for research and problem solving, Seminar Handbook, The Welding Institute, Cambridge U.K., pp 21-23, 1972
6. Ion J.C. and Easterling K.E. Proceedings of Third Scandanavian Symposium in Material Science, Oslo Finland, pp 79-85
7. Rosenthal D. The theory of moving sources of heat and its application to metal treatments, Trans ASME, Vol. 68, pp 849-865, 1946
8. Rykalin R.R. Energy sources for welding, Welding in the World, Vol. 12, No. 9/10, p 227-248, 1974
9. Ion J.C., Easterling K.E. and Ashby M.F. A second report on diagram of microstructure and hardness for heat affected zones in welds, Acta Metallurgica, Vol. 32, pp 1949-1962, 1982
10. Easterling K.E., Ashby M.F. and Li. A first report on diagrams for grain growth in welds, Acta Metallurgica, Vol. 30, pp 1969-1978, 1982
11. Bastien P., Drollet J. and Maynier P. Prediction of microstructure via empirical formulae based on CCT diagrams, Hardenability Concepts with an Application to Steel, pp 163-176, 1977
12. Arata Y., Matsuda F. and Nakata K. Japanese Welding Research Institute Journal, Vol. 1, p 39, 1970
13. Bibby M. J., Chong L.M. and Goldak J.A. Predicting heat affected zone hardness by the weld test method, Journal of Testing and Evaluation, ASTM, Vol. 11, p 126, 1983
14. Alberry P.J. and Jones W.K.C. Computer model for prediction of heat affected zone microstructure in multipass weldments, Metal Technology, Vol. 9, pp 419-426, Oct. 1982
15. Rosenthal D. The theory of moving sources of heat and its application to metal treatments, Trans ASME, Vol. 68, pp 849-865, 1946
16. Kirkaldy J.S. and Venugopalan: Prediction of microstructure and hardenability in low alloy steels, phase transformation in ferrous alloys, Proceedings of the International Conference, Oct. 4-6, 1983
17. Kirkaldy J.S. and Sharm R.C. A new phenomenology for steel IT and CCT Curves, Scripta Metallurgical, Vol. 16, p 1193, 1982
18. Kirkaldy J.S. Prediction of alloy hardenability from thermodynamic and kinetic data, Metallurgical Transactions, Vol. 4, p 2327, 1973
19. Kirkaldy J.S. and Baganis E.A. Thermodynamic prediction of the Ae3 temperature of steels with additions of Mn, Si, Cr, Mo and Cu, Metallurgical Transactions, Vol. 9A, pp 495-501, 1978
20. Henwood C.E. (1998). An analytical model for computing weld microstructure. Master's thesis, Carleton University.
21. Watt D. F., Coon L., Bibby M. J., Goldak J.A. and Henwood C. An algorithm for modeling microstructural development in weld heat affected zones, Acta Metal, Vol. 36, No 11, pp 3029-3035, 1988

22. Henwood C. Bibby M.J., Goldak J.A. and Watt D.F. Coupled transient heat transfer microstructure weld computations, Acta Metal, Vol. 36, No. 11, pp 3037-3046, 1988
23. Gu M. (1992). Computational weld analysis for long welds. Doctoral thesis Carleton University.
24. Lohner R. Commn. Appl. Numer. Meth., Vol. 4, p 123, 1988
25. Gu M., Goldak J.A. and Hughes E. Modeling the evolution of microstructure in the heat-affected-zone of steady state welds, Can. Metall. Quarterly 32, No. 4, pp 351-361, 1993
26. Goldak J., Chakravarti A. and Bibby M. A finite element model for welding heat sources, Metallurgical Transactions B, Vol. 15B, pp 299-305, June 1984
27. Bibby M.J., Eastman K. and Goldak J.A. Metallurgical Transactions, Vol. 14B, p 483, 1983
28. Cameron S. IEEE Computational Graphics Applications, Vol. 68, May 1991
29. Strang G. Introduction to applied Mathematics, Wellesley Cambridge Press, Wellesly Massachusetts USA, 1986
30. Gu M., Goldak J.A. and Bibby M.J.; Computational heat transfer in welds with complex weld pool shapes, Adv. Manuf. Eng., Vol. 3. January 1991
31. Barlow J.A. One weld or two? The formation of a submergend arc weld pool, Welding Inst. Res. Bull., pp 177-180, June 1982
32. McKellinget J. and Szekely J. Metall. Trans., Vol. 17A, p 1139, 1986
33. Leslie W.C. The physical metallurgy of steels, McGraw-Hill Book Company, pp 256-276, 1981
34. Andrews K.W. Journal of Iron Steel Institute, Vol. 203, p 721, 1965
35. Lu W.K. McMaster University, Private Communication to D.F. Watt.
36. Ion J. C. (1984). Modeling of microstructural changes in steels due to fusion welding, Doctoral Thesis, Lulea University.
37. Maynier Ph. and Drollet J. Creusot-Loire system for the prediction of the mechanical properties of low alloy steel products, hardenability concepts with application to steels, Trans. AIME, p 518, 1977
38. Adams C.M. Cooling rate and peak temperatures in fusion welding, Welding Journal AWS, Vol. 37, No. 5, p 210, 1958

Chapter V

Evolution of Microstructure Depending on Deformations

5.1 Introduction and Synopsis

The thermal cycle imposed on any welded object causes thermal expansions and contractions to occur that vary with time and location. Since this expansion is not uniform, stresses that appear in hot regions near the weld are restrained by cooler regions further away. Plastic deformation, occurring as a result of these stresses, leads to residual stresses in the object that remain after the temperatures have returned to ambient levels. Thus, the use of welding in the fabrication or repair of pressure vessels introduces residual stresses in the weld itself and in the base metal. An accurate knowledge of the magnitude and distribution of the residual stresses is useful when assessing the fitness of the vessel.

Three fundamental dimensional changes of a flat welded plate are:

- transverse shrinkage;
- longitudinal shrinkage;
- angular distortion (rotation around the weld line)

From a more detailed point of view, the welding deformations can be classified as [48], Figure 5-1:

- transverse shrinkage - shrinkage perpendicular to the weld centerline;

- longitudinal shrinkage - shrinkage in the direction of the weld line;
- angular distortion-distortion caused by nonuniform temperature distribution in the through-thickness direction;
- rotational distortion-angular distortion in the plane of the plate due to thermal expansion or contraction;
- bending distortion-distortion in a plane through the weld line and perpendicular to the plate;
- buckling-distortion caused by compressive stresses inducing instability particularly when the plates are thin

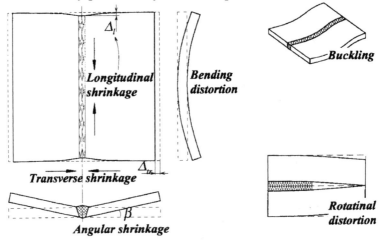

Figure 5-1: Various types of welding distortion, from [48].

Material modeling is, together with the uncertainty in the net heat input, one of the major problems in welding simulation [1 and 2]. The thermal analysis is, more straightforward than the mechanical analysis. It entails few numerical problems, with the exception of the large latent heat during the solid-liquid transition, and it is easier to obtain accurate values of the thermal properties rather than the mechanical properties of a solid.

McDill et al [3] investigated the relative importance of the thermal and mechanical properties of stainless and carbon steels in welding simulations. Two bars of dimension *20in x 2in x 0.5in* were "welded" along the free edge. No fixture was used. One bar was made of stainless steel and the other of carbon steel. The resultant radius of curvature was compared with the computed values for the material properties of carbon steel (MS) or stainless steel (SS). The results are shown in Table 5-1. The somewhat unexpected conclusion was that the thermal properties play a more important role than the mechanical properties in explaining the different behavior between these steels. This is due to the fact that the thermal dilatation is the driving force in the deformation and the bar is free to bend. The thermal dilatation is determined by the temperature field and is, therefore, strongly influenced by the thermal properties. In other words, the thermal dilatation together with the temperature distribution is important for the deformation [1].

Table 5-1: Curvature for combinations of thermal and mechanical properties [3]. MS mild steel, SS 304 stainless steel

Test	Thermal Properties	Mechanical Properties	Radius of Curvature (1/m)
A	MS	MS	41.6
B	SS	SS	11.9
C	MS	SS	29.6
D	SS	MS	16.3
Experiment	MS	MS	23.3
Experiment	SS	SS	8.1

The complete thermo-mechanical history of a material will influence its material properties. However, in many cases an adequate approximation is to make properties a function only of the current temperature and effective plastic strain.

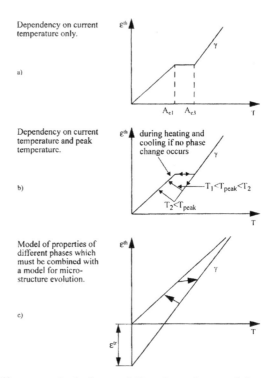

Figure 5-2: Different methods for modeling dependency of thermal dilatation on phase changes in ferritic steel, from Lindgren [1].

The simplest and most common approach is to ignore the microstructure change and assume that the material properties depend only on temperature. Then, for example, one might assume that the thermal dilatation is the same during heating and cooling, as in Figure 5-2a.

Distortion and residual stress in welds are caused by nonlinear material behavior. For sufficiently low stresses and sufficiently short times, the stress strain relationship for solids is linear elastic. For sufficiently high stresses and sufficiently long times, the stress strain relationship is nonlinear. What a sufficiently low stress and a short time is, depends the problem, its material type, its microstructure and the temperature.

For Newtonian fluids the stress strain rate relationship is linear. The nonlinear behavior of a solid tends to be strain rate dependent at high temperatures and sufficiently long times. This means that the stress is a function of the strain rate. The plastic deformation is often associated with the thermally activated motion of dislocations in the presence of a deviatoric stress field. This is called rate dependent plasticity. At low temperatures, the strain rate is a nonlinear function of stress rate and is not a function of stress.

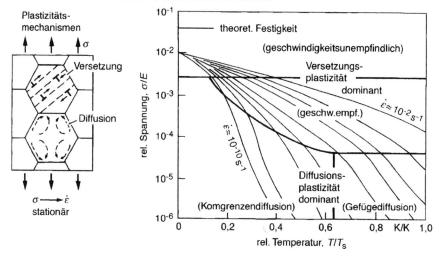

Figure 5-3: Sketch of an Ashby diagram illustrating deformation mechanisms and different strain rates. Different deformation mechanisms in different temperature and stress regimes are shown, from [43 and 44].

To accurately predict stress, residual stress and distortion in welded structures, one must know how the material responds to a given stress or strain, i.e., which constitutive equation to use for a given state of the material and loading condition. One must know which constitutive equation is most appropriate as well as the material properties that describe this behavior such as yield stress or deformation resistance. The best known representation of material behavior versus strain rate and temperature are Ashby diagrams, Figure 5-3.

The mechanisms for the plastic deformation vary due to the large temperature range involved. In welds, the strain rate is usually less than *0.01/second*. A crack driven by elastic unloading could have much higher strain rates but that is not considered here. The Ashby diagram has several striking features. First there are many different mechanisms for deformation. Therefore one should not expect a simple function. Second the flow stress rises rapidly with decreasing temperature. Third the difference in flow stress for strain rates of say *0.0001/s* and *0.01/s* narrows rapidly as temperature is decreased.

The homologous temperature is defined as the current temperature in Kelvin divided by the melting temperature in Kelvin. The authors recommend that a linear viscous model be used at a homologous temperature above *0.8*. Rate dependent plasticity be used down to a homologous temperature of *0.5* and rate independent (usually with von Mises plasticity), [51], for lower temperatures.

5.2 Properties for Modeling

Accurate modeling of the development of residual stresses and strains in welded steels requires realistic modeling of the microstructural evolution. It is possible to obtain accuracy comparable to experimental measurements.

In the equations of continuum mechanics, the parameters such as thermal conductivity, elasticity tensor, the internal variables controlling plasticity or visco-plasticity are sensitive functions of microstructure. Volumetric strain rates due to phase changes and thermal expansion are the dominate loads in the stress analysis of welds. Although latent heats of solid-solid transformations do not have a large affect, they can be detected. If the weld pool shape is not taken as data, the evolution of the latent heat of fusion has a large affect in the energy equation. Solidification kinetics strongly influences the weld pool shape. Certainly, the equations for Magneto-hydro-dynamics *(MHD)* in the arc and the Navier-Stokes in the weld pool are much more difficult to solve than the continuum mechanics of the solid.

FEM thermal stress analyses of welds began with Ueda in *1972*. In the *1980s* this capability was extended significantly. With few exceptions, these analyses were Lagrangian, i.e., the *FEM* mesh was tied to the material. In continuum mechanics, there are two basic descriptions, resulting in two basic formulations, the Lagrangian formulation and the Eulerian formulation.

The Eulerian formulation uses spatial coordinates. With the Eulerian formulation material points move through the observed domain, which is fixed in space. The Lagrangian formulation uses material coordinates so that the observed domain deforms and no material moves into or out of the domain. The Lagrangian formulation is often chosen when the constitutive parameters in the deformation process are history dependent and the deformed geometry is unknown in advance. The Eulerian formulation is preferred when the deformed geometry is known in advance and the process is history independent and at steady state.

The arbitrary Eulerian Lagrangian formulation used initially in fluid mechanics [8 and 9] and its twin in solid mechanics, the mixed Eulerian Lagrangian formulation of Haber [10], and Koh et al. [11], has the capability to solve history independent problems with evolving fields, in which there is often a moving patch with a steady state field. No history dependent variables are involved in Haber's formulation. Ghosh and Kikuchi [12] published a paper on the arbitrary Eulerian Lagrangian formulation with the history dependent variables interpolated from a pseudo-material mesh. These results added computational cost and a possible loss in accuracy due to the interpolation across elements.

In the Eulerian formulation for steady state stress analysis with history dependence, the history is often embedded in flow lines. The flow lines are formed by the method of characteristics starting from a known inlet boundary. This was done by Dawson [13] for the rolling process, in which the deformed or current boundary locations must be known.

In an arbitrary Eulerian Lagrangian *(AEL)* formulation, the mesh is allowed to move with the material points or with spatial points or allowed any other convenient mesh motion or velocity. In this case, the task of computing flow lines or streamlines is not fundamentally different from the Eulerian case. One simply deals with both the velocity of material points and the velocity of the mesh.

The steady state formulation proposed by Gu and Goldak up to 1995 [6, 28 and 29] applies especially to stress analysis with history dependence and with unknown deformed boundary, particularly in weld processes. The reference mesh and the arc are fixed in space and do not move. The material moves relative to the deformed mesh at the welding speed. At the same time every node in the frame moves continuously with the deformation.

In the past two decades considerable progress has been achieved in analyzing the mechanics of welds in the liquid region using the Navier-Stokes equations and in the solid region at lower temperatures using rate independent plasticity with hardening. Progress has been slow in stress analysis in the solid region near the melting point that must deal with both the rate dependent plasticity when a material point is above $0.5T_m$ and with rate independent plasticity when a material point is below $0.5T_m$.

At low temperatures roughly below $0.5T_m$, the strain rate is a nonlinear function of stress rate. To accurately predict stress, residual stress and distortion in welded structures, one must know how the material responds to a given stress or strain, i.e., which constitutive equation to use for a given state of the material and loading condition. Not only must one know which constitutive equation is most appropriate, one must know the values of material properties, i.e., the internal variables that describe this behavior such as yield stress or deformation resistance.

In this model, four interacting agents are created to solve the coupled problem. These agents are Energy, Viscosity, Material Properties and Stress. They are called in this order in each time step.

At any particular time step each finite element implements one of the three models depending on the average temperature of the element. Thus the constitutive model is element level data and all Gauss points in an element implement the same constitutive for that time step.

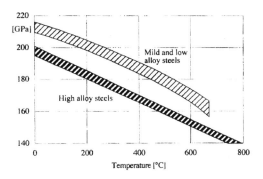

Figure 5-4: Young's modulus for some steels, from Richter [34 and1].

Young's modulus and Poisson's ratio are shown in Figures 5-4 and 5-5 for lower temperatures [34].

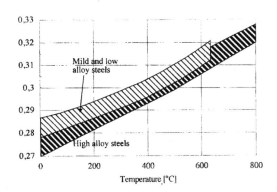

Figure 5-5: Poisson's ratio modulus for some steels, from Richter [34 and 1].

The temperature in welds varies from the boiling temperature to room temperature. The domain includes the liquid weld pool and the far field of solid near room temperature. The liquid is well described as a Newtonian fluid characterized by a temperature dependent viscosity. The temperature of solid ranges from the melting temperature T_m to room temperature, where T_m is in degrees of Kelvin. At temperatures from 0.8 to $1.0 T_m$, the solid can be considered to be a linear viscous material characterized by a viscosity due to the diffusion of dislocations. From temperatures of 0.5 to $0.8 T_m$, the solid can be considered to be viscoplastic and characterized by an elasticity tensor, viscosity and deformation resistance. Below $0.5T_m$, the solid can be considered to be a rate independent plastic material characterized by an elasticity tensor, yield strength and isotropic hardening modulus. Also three basic classes of constitutive behavior (three plastic potential functions) for low, intermediate and high temperature plasticity can be defined. All models were implemented in the same Lagrangian finite strain frame work [33]. The constitutive model was assigned to be rate independent if the temperature is less than $0.5T_m$, rate dependent if the temperature is in a range 0.5-0.8 T_m and linear viscous if the temperature was greater than $0.8T_m$. Thus the constitutive equation of a material point had to be able to change type with space and time during the welding process. The results of the analysis in [33] agreed in a qualitative sense with the experimental results of Chihoski [47].

The rate dependent plasticity constitutive equation is only applied on the sides of the *HAZ*. In front of the weld pool, the temperature rise is so rapid that this mesh is too coarse to capture the thin rate dependent plastic zone. Note that the pressure is highest in this zone. Also note that the cooling weldment pulls metal back towards the weld pool. The effective plastic strain reaches a maximum some distance behind the weld pool.

5.2.1 Stresses, Strains and Deformations

Thermal-elastic-plastic constitutive models decompose the total strain rate $\dot{\varepsilon}_{ij}^{Tot}$ into the elastic $\dot{\varepsilon}_{ij}^{e}$, plastic $\dot{\varepsilon}_{ij}^{p}$ due to rate independent plasticity, thermal $\dot{\varepsilon}_{ij}^{th}$ consisting of thermal expansion and creep strain rate $\dot{\varepsilon}_{ij}^{c}$. During phase transformations additional terms, $\dot{\varepsilon}_{ij}^{Trv}$, i.e., the strain rate volume change associated with the transformation and $\dot{\varepsilon}_{ij}^{Trp}$, i.e., the strain rate transformation plasticity are included:

$$\dot{\varepsilon}_{ij}^{Tot} = \dot{\varepsilon}_{ij}^{e} + \dot{\varepsilon}_{ij}^{th} + \dot{\varepsilon}_{ij}^{p} + \dot{\varepsilon}_{ij}^{c} + \dot{\varepsilon}_{ij}^{Trv} + \dot{\varepsilon}_{ij}^{Trp} \qquad (5\text{-}1)$$

The volume change associated with the transformation is often included as part of the thermal strain by modifying the coefficient of thermal expansion. Separation of the transformation volume change from the thermal strain allows more flexibility in the description of the material [39]. Oddy et al. [40] defined $\dot{\varepsilon}_{ij}^{Trp}$ to be a function of the stress deviator T', relative transformation volume change $\Delta V/V$, transforming phase fraction z and an internal variable s:

$$\dot{\varepsilon}_{ij}^{Trp} = \frac{5}{4}\frac{\Delta V}{V}\frac{1}{s}T'(2-z)z \qquad (5\text{-}2)$$

where the internal variable s is a yield stress as in [41] or deformation resistance as in [33] and T' is the deviatoric stress tensor.

The incremental constitutive models used with temperature dependent material properties require the increment in the strain that occurs during the time step. Thus, for a step from time n to time n+1, the incremental form of (5-2) is:

$$\Delta\varepsilon_{ij}^{Trp} = \frac{5}{4}\frac{\Delta V}{V}\frac{1}{s}T'(2-2z_{(n)}-\Delta z)\Delta z \qquad (5\text{-}3)$$

where $z_{(n)}$ is the fraction already transformed at the start of the increment and Δz is the amount which transforms during the increment. A more accurate method of evaluating the incremental strain is by integrating the appropriate strain rate over the increment. Equation (5-3) can also be obtained from the time derivative of (5-2), integrated over the time step using the midpoint value, assuming a piecewise constant stress.

The effect of different material models on the residual stresses is shown in Figure 2-27 for conditions resembling the case studied by Hibbitt and Marcal [49]. Oddy et al [39] gives a concise description of transformation-induced plasticity *(TRIP)*. Figure 5-6 shows the through-thickness distribution of the longitudinal (axial residual stresses on the plane of symmetry), at the mid-length of the weld, with and without transformation plasticity. A significant variation in the longitudinal stress through the thickness is evident. In both cases, the highest stress exists on the inner surface. Although the curves are similar in shape, with and without transformation plasticity, the magnitudes are significantly different.

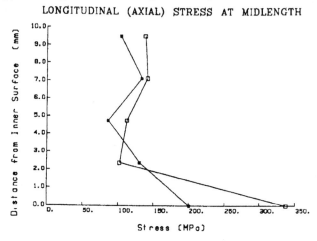

Figure 5-6: Residual longitudinal stress versus distance from inner surface, at midlength ──▪ including transformation plasticity, ──◦ excluding transformation plasticity, from Oddy et al [39].

The magnitude of residual stresses present after welding is important in the prediction of the resistance to fracture. In this case the transverse or hoop stresses are of particular interest because they will influence the growth of axial flaws.

Figure 5-7 compares the through-thickness variation of the midlength of the weld. The transverse stress is much smaller than the longitudinal stress. In both cases the largest values are found near, but not at the inner surface. The inclusion of transformation plasticity has made a substantial difference in the transverse stress predicted on the inner surface. The magnitude of the residual transverse stress increased when transformation plasticity was included.

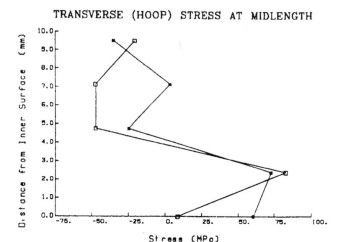

Figure 5-7: Residual transverse stress versus distance from inner surface, at midlength —■ including transformation plasticity, —□ excluding transformation plasticity, from Oddy et al [39].

The stress-dependence of the microstructure evolution is central to transformation induced plasticity *(TRIP)* in martensitic transformations. Transformation plasticity arises from 2 sources. The transformation creates local, microscopic stresses and plastic strains. The interaction of these microscopic fields with the global, macroscopic stress fields results in observable, irreversible, macroscopic plastic deformation known as the Greenwood-Johnson effect. The stress fields also influence the orientation of the shear and dilatational deformations of martensite formation. This leads to the Magee effect [42].

To specify a particular constitutive model we must specify the plastic potential function (or dissipation function) Φ, (5-4), and the evolution equation for state variables, (5-5):

$$\Phi = \frac{\dot{\varepsilon}_0 s}{(\frac{1}{m}+1)} \{\frac{\tilde{\sigma}}{s}\}^{[(\frac{1}{m}+1)]} \tag{5-4}$$

wrt effective stress $\tilde{\sigma}$ in the limit of very low rate sensitivity m.

$$\dot{s} = hD^p - \dot{r} \tag{5-5}$$

where the function h governs the rate of strain hardening and dynamic softening, and the function $\dot{r}$ governs the rate of static recovery. D^p is the plastic deformation rate tensor.

The deformation gradient caused by elastic and plastic strain is defined as follows:

$$F = F^{Th^{-1}}(I + \nabla u) \qquad (5\text{-}6)$$

where u is the displacement vector, ∇u is the displacement gradient and F^{Th} is the thermal deformation gradient. Superscript $^{-1}$ denotes the inverse of the matrix. I is the 2nd order identity tensor. In the simplest case of thermal-mechanical coupling it becomes:

$$F^{Th} = \alpha \cdot \Delta T \cdot I \qquad (5\text{-}7)$$

ΔT is the temperature increment for the time step and α is the thermal expansion coefficient.

The total rate of stretching is:

$$D = D^e + D^p \qquad (5\text{-}8)$$

Superscripts e and p denote the elastic and plastic parts of the rate of stretching, respectively.

Suppose we have a set of stress and state variables at time $t\{\sigma,\ S\}$ and we also have an estimate or trial value for the same set at time $\tau = t + \Delta t$. Then a stable, accurate and efficient algorithm to compute the set of stress and state variables at time τ is needed. Also a linearization of the constitutive equation is needed to compute the element stiffness that is assembled into a global stiffness matrix. The consistency of this linearization is very important to achieve robust and rapid convergence of the global linearization of momentum equation.

Assume that both the plastic and total stretching rate are constant within the time step. The time derivative of Hooke's law [33 and 35] and Goldak et al [7] yields:

$$\bar{\sigma}(\tau) = \sigma(t) + \varsigma[\int Dd(\xi) - \int D^p d(\xi)]_t^\tau \qquad (5\text{-}9)$$

and for the state variables we arrive at:

$$S(\tau) = S(t) + \int_t^\tau \dot{S} d(\xi) \qquad (5\text{-}10)$$

We assume again that $\dot{S}$ is constant over the time step. Where:

ς = elasticity tensor, 4th order

σ = Cauchy stress

$\overline{\sigma}$ = convected Cauchy stress

ξ = dimensionless activation volume for dislocation diffusion

In every constitutive model the set of Equations (5-9) and (5-10) is a system of nonlinear equations that must be solved for each time step, at each integration point of the space domain for each global iteration, usually using a Newton-Raphson iteration.

The isotropic constitutive model suggests that for a particular integration point the elasticity tensor ς is a linear combination of the two independent elastic constants:

$$\varsigma = 2\mu\Re + (\kappa - (2/3)\mu)1 \otimes 1 \qquad (5\text{-}11)$$

where μ and κ are the elastic shear and bulk modulus, $\Re$ and 1 are the fourth and second-order identity tensors, respectively.

The two following cases of rate independent plasticity and linear viscous plasticity are obtained from the plastic potential function (5-4) by substituting rate sensitivity $m \rightarrow 0$ and $m = 1$, respectively. The $m \rightarrow 0$ means that $(1/m)$ is very large compared to 1 in the exponent in (5-4).

In the welding process, changes in stress caused by deformation are assumed to travel slowly compared to the speed of sound. So, at any instant, an observed group of material particles is approximately in static equilibrium, i.e.: inertial forces are neglected. In rate independent plasticity, viscosity is zero and viscous forces are zero. In either the Lagrangian or the Eulerian reference frame, the partial differential equation of equilibrium is, at any moment:

$$\frac{\partial \sigma_{ji}}{\partial x_j} + f_i = 0 \quad \text{or} \quad \nabla \cdot \sigma + \mathbf{f} = 0 \qquad (5\text{-}12)$$

where $\mathbf{f}$ is the sum of the body forces and σ is the Cauchy stress.

In the *FEM* formulation, Equation (5-12) is transformed and integrated over the physical domain, or a reference domain with a unique mapping to the physical domain [6].

5.2.2 Rate Independent Isotropic Plasticity

Rate independent plasticity occurs at low temperatures, roughly in the temperature range below $0.5T_m$. The deformation is due to dislocation glide and strain rate due to thermal fluctuations plays no significant role. The relaxation time is zero. In rate independent plasticity the total strain rate is decomposed into elastic and plastic strain rate.

Introduce the identities:

$$\dot{\tilde{\varepsilon}}^p \equiv \sqrt{(2/3)D^p D^p} \geq 0 \qquad (5\text{-}13)$$

is the effective plastic strain rate or equivalent plastic strain rate.

The plastic stretching or deformation rate tensor is:

$$D^p \equiv \sqrt{3/2}\,\dot{\tilde{\varepsilon}}^p N^p \qquad (5\text{-}14)$$

where N^p is the unit tensor defining the direction of plastic flow.

Introduce the outward unit normal to the yield surface at the current stress point:

$$N \equiv \sqrt{3/2}\,T' / \tilde{\sigma} \qquad (5\text{-}15)$$

At this point we adopt the classic associated plasticity (or normality) flow rule:

$$N^p = N \qquad (5\text{-}16)$$

The rule (5-16) means that the principal axes of plastic stretching are the same as the principal axes of deviatoric stress or, in other words, the direction of plastic flow is the outward normal to the yield surface at the current stress point.

Consider the derivative of the plastic potential function (5-4) with respect to effective stress $\tilde{\sigma}$:

$$\dot{\tilde{\varepsilon}}^p = \dot{\varepsilon}_0 \{\frac{\tilde{\sigma}}{s}\}^{1/m} \qquad (5\text{-}17)$$

The yield phenomenon suggests that there is a switching from $\dot{\tilde{\varepsilon}}^p = 0$ if $\tilde{\sigma} < s$ to an unbounded plastic strain rate if $\tilde{\sigma} > s$. That is the case if $m \rightarrow 0$. Physically this means that an unbounded plastic strain rate immediately leads to material hardening (for a hardening material) that, in turn, moves the yield surface up to the stress point and satisfies the constraint $s = \tilde{\sigma}$ at any time moment. Then define the evolution equations for deformation resistance (isotropic hardening):

$$\dot{s} = h\dot{\tilde{\varepsilon}}^{\,p} \qquad (5\text{-}18)$$

For rate independent isotropic plasticity, the time integration of (5-9) and (5-10) is implemented by [35]. Full details are presented in [7].

Steady-State Formulation

It has long been recognized that a quasi-steady state temperature field exists in certain long welds. If only thermal loads from a weld heat source are present in such welds, due to this quasi-steady state, all materials points on one flow line or trajectory in these welds will experience the same thermal cycle. Microstructures and mechanical properties in a weldment are expected to change along flow lines during welding. These observations have motivated the analysis of stresses in such steady state welds.

The notable points of steady state formulation for stress analyses of welds are the following:

-The kinematic model differs from the usual Lagrangian or Eulerian models.

-History-related quantities at a material point are obtained from streamlines in a reference frame after some mapping.

-The solution includes history dependent effects such as plasticity and microstructure evolution. It includes also the strain caused by phase transformation.

-Tests have been done on an edge weld. The distortion in the welded bar is clearly demonstrated.

-Satisfactory results are obtained. Computed longitudinal stresses have been compared with the data calculated in the Lagrangian formulation and measured in experiments.

-The convergence rate is similar to that of a Lagrangian formulation.

-Significant savings of computing time and memory usage have been achieved.

This method has several distinguishing features:

(a) The thermal strain creates the load in the weldment. If there is no constraint on a weldment, it is the only load.

(b) The mechanical properties of most materials are both temperature and (plasticity and microstructure) history dependent

(c) Heat sources, and therefore the loads, are remarkably localized.

(d) The resulting thermal stress is high enough to cause plastic deformation.

(e) Phase transformations can either increase or release stresses.

The steady state analysis differs from the usual Lagrangian formulation for calculating the stiffness matrix and residual vector in that the current formulation takes Gauss point data from flow lines for computation. Every Gauss points knows the element it resides in, which makes it easier for a Gauss point to make contributions to the element, or to the nodes of the element. Since the Gauss points are object-oriented, pointers from Gauss points to elements are simple and natural.

The steady state formulation uses flow lines to trace the history and thus the evolution of all internal variables. Before welding, the mesh has no distortion, and the elements of the mesh are regular, Figure 5-11. The Gauss points in elements are aligned to form flow lines by the methods and data structure described by Gu M. in references [6, 25 and 50].

The Gauss points are initially aligned to form straight lines in the undeformed state. Since the mesh deforms with respect to a Lagrangian frame, these straight lines distort continuously with the deformation. They are the exact flow lines for the current state when the convergence is reached. The whole nonlinear approach is to compute these flow lines.

Constitutive equations describe the material response to the strains at the current state. In a numerical approach, it is customary to assume initially that the deformation is elastic. In Gu et al [6], the elasto-plastic constitutive model uses a modified effective-stress-function *(ESF)* algorithm [15, 16 and 17].

The trial stress, $^i\sigma^*$, at each integration point is assumed to be:

$$^i\sigma^* = {}^{i-1}\sigma + \int_{i-1}^{i} Dd\varepsilon \qquad (5\text{-}19)$$

where D is a modified material property tensor. The superscript i indicates a material point x (at time t) on flow line containing a spatial point x.

The $i\text{-}1$ indicates the same material point at time $(t - \varDelta t)$ on the flow line. The incremental trial stress, therefore,

$$\varDelta^i\sigma^* = \int_{i-1}^{i} Dd\varepsilon \qquad (5\text{-}20)$$

Including thermal, elastic, plastic, transformation and creep deformation, constitutive equations can be written in the form:

$$^iT = \frac{^iE}{1 + {}^iv}({}^ie - {}^i\varepsilon^p - {}^i\varepsilon^c - {}^i\varepsilon^{Trp}) \qquad (5\text{-}21)$$

$$^i\sigma_m = \frac{^iE}{1 - 2{}^iv}({}^i\varepsilon_m - {}^i\varepsilon^{th} - {}^i\varepsilon^{Trv}) \qquad (5\text{-}22)$$

where at point i:

^iT = deviatoric stress tensor
   = $^i\sigma - {}^i\sigma_m$

^ie = deviatoric strain tensor
   = $^i\varepsilon - {}^i\varepsilon_m$

$^i\varepsilon^p$ = plastic strain tensor

$^i\varepsilon^c$ = creep strain tensor

$^i\varepsilon^{Trp}$ = transformation plastic strain tensor

$^i\sigma_m$ = stress tensor

$^i\varepsilon_m$ = strain tensor

$^i\varepsilon^{th}$ = thermal strain tensor

$^i\varepsilon^{Trv}$ = strain due to volume change in phase transformation

$^iE, {}^iv$ = Young's modulus and Poisson's ratio at point i corresponding to temperature at point i

The *ESF* algorithm has been adopted to increase the efficiency of solving for the incremental plastic strain, $\Delta\varepsilon^p$, as well as the incremental creep strain, $\Delta\varepsilon^c$. A brief description of *ESF* used in the current case is given by Gu et al [6 and 50]. Details of the *ESF* can be found in Kojic and Bathe [9 and 10]. The transformation plasticity formulation is due to Oddy et al [17 and 42].

Deformation

By introducing a reference configuration in the deformation equations, any configuration in the deformation process can be expressed in that reference configuration. Figure 5-8 shows the displacement of a particle from its initial position ^{i-1}x to the current position $^i x$. The particle's reference position can be related to both deformed positions through a motion:

$$x = x(^r x, t) \tag{5-23}$$

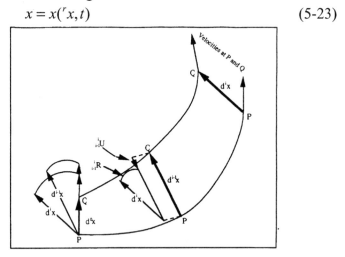

Figure 5-8: Displacement, stretch and rotation of a vector $d^{i-1}x$ at position i-1 to $d^i x$ at a new position i. Both have unique mappings to the reference configuration, from [6].

The deformation gradient tensor is the simplest to define in terms of the deformation equations. This tensor also includes more information than the strain tensor. R and U can be obtained by a polar decomposition of a relative deformation gradient tensor [6, 21,

22 and 23]. A polar decomposition $F=R\ U$ describes both the rotation R and stretching U experienced during the deformation.

The deformation gradient ^iF maps vectors dx^r to vectors d^ix. Hence, in this context ^iF is defined as the tensor whose rectangular Cartesian components are the partial derivatives $\partial^ix_k/\partial^rx_m$. The tensor ^iF operates on an arbitrary infinitesimal material vector d^rx at ^rx. It associates the vectors d^ix and d^rx as:

$$d^ix=^iF\cdot d^rx \qquad (5\text{-}24)$$

$$d^ix_k=\frac{\partial^ix_k}{\partial^rx_m}d^rx_m \qquad (5\text{-}25)$$

One advantage of the Lagrangian formulation is that it traces the history of changes of mechanical properties in the deformed material naturally. In the updated Lagrangian formulation the mechanical properties are updated in every time step. At the beginning of every step, all history results are treated as initial conditions. The formulation deals with the same material points in the final and initial configurations. Without deformation, a local coordinate system will remain unchanged in two configurations. An infinitesimal vector is computed in the local coordinate system.

When the deformation occurs between two configurations, the local coordinate system can change its length and direction. So does the infinitesimal vector. In general, this local coordinate system differs from one point to another in a curvilinear coordinate system. Each element would no longer contain the same set of material points. The stress or strain history of deformation at a material point can only be found from an upstream position of a flow line. The position is currently taken by another material point and the local coordinate system at that position is unlikely to be the same one as that of the current point, even without deformation. A mapping strategy is necessary.

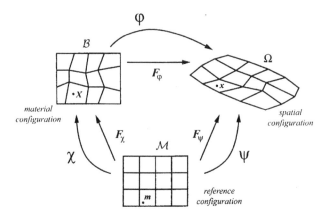

Figure 5-9: The *AEL* kinematics. The physical motion $\varphi := {}_{i}^{i-1}R$ of the solid or fluid is decomposed in the material motion $\chi := {}^{i-1}R$ and the mesh motion $\psi := {}^{i}R$ for a reference mesh M , adopted from [36].

Figure 5-9 indicates the relationships between the two deformed configurations and the undeformed configuration (reference configuration). By choosing a reference configuration and by mapping material points in any of the other configurations to the reference configuration, it is possible to obtain the deformation gradient tensor between any two deformed configurations. A few iterations are required to reach this indirectly from the deformed configurations to the reference configuration or in reverse order.

A relative deformation gradient tensor, ${}_{i-1}^{i}F$, between initial configuration i- 1 and final configuration i, can be defined:

$${}_{i-1}^{i}F(x) = \frac{\partial^{i}x}{\partial^{i-1}x} = \frac{\partial^{i}x}{\partial^{r}x}\frac{\partial^{r}x}{\partial^{i-1}x} = {}^{i}F^{i-1}F^{-1} \tag{5-26}$$

In indicial form:

$${}_{i-1}^{i}F_{mn} = \frac{\partial^{i}x_{m}\partial^{r}x_{k}}{\partial^{r}x_{k}\partial^{i-1}x_{n}} \tag{5-27}$$

Strain Increment Measurement

The stress in the current deformed configuration is called the Cauchy stress. If the current deformed configuration is mapped to any other configuration such as a reference configuration in AEL (arbitrary Eulerian Lagrangian) or the configuration at the beginning of a time step, the Cauchy stress must be mapped to this configuration. Different mapping methods result in different stress types. The 2nd Piola-Kirchhoff, see chapter *VI*, and 1st Piola-Kirchhoff are examples. Each stress has its conjugate strain type defined such that the product $\sigma{:}\varepsilon$ equals the strain energy density. For each stress-strain type, the constitutive equation must be defined. In the following, when the stress rate is specified, the corresponding strain rate is implied.

For a weld, in some local area within the heat affected zone, the displacement-gradient components can be of the order of unity. Thus, regions far from the heat affected zone may have large displacements and large rigid body rotations but small strains. Looking at a weldment, a large rigid body rotation from the original shape can often be seen with the naked eye. Infinitesimal strain measures can cause large errors in the residual stress analysis for welds. In such situations, finite strain measures should be used for weld analyses.

Many research papers have exploited various stress rates. The advantages and disadvantages of different stress rate formulations are described in references [18, 19, 20 and 21].

Along a flow line, the incremental logarithmic strain at point "*i*" is:

$$\Delta\varepsilon = {}_{i-1}^{i}R\ln({}_{i-1}^{i}U){}_{i-1}^{i}R^{T} \tag{5-28}$$

where ${}_{i-1}^{i}R$ is a rotation tensor rotating the stress or strain in the orientation of point "*i-1*" to that of the point "*i*", and ${}_{i-1}^{i}U$ is a stretch tensor describing the stretch from "*i-1*" to "*i*" in the orientation of stress (or strain) at "*i-1*".

With Equations (5-21), (5-22), (5-26) and (5-27) a summary of the current formulation to calculate incremental stress from point "*i-1*" to "*i*" along a flow line is as follows, from Gu et al [6]:

Step 1: Decompose the relative deformation gradient at point "*i*",

$$\tag{5-29} {}_{i-1}^{i}F = {}_{i-1}^{i}R{}_{i-1}^{i}U$$

Step 2: Calculate the incremental logarithmic strains at the strain orientation of point "i-1",

$$\Delta\varepsilon' = \ln(_{i-1}^{i}U)$$

$$\Delta\varepsilon_m = \frac{tr(\Delta\varepsilon)}{3} \qquad\qquad (5\text{-}30)$$

$$\Delta e = \Delta\varepsilon' - \Delta\varepsilon_m$$

where Δe is the increment in the deviatoric strain tensor from i-1 to i.

Step 3: Calculate the total Cauchy stresses at the stress orientation of point "i-1",

$$^{i-1}(^{i}\sigma) = {^{i-1}}\sigma + \int_{i-1}^{i} Dd\varepsilon$$

$$^{i}\sigma = \frac{^{i}E}{1+^{i}v}(^{i-1}e^e + \Delta e - \Delta\varepsilon^p - \Delta\varepsilon^c - \Delta\varepsilon^{Trp})$$

$$= \frac{^{i}E}{1+^{i}v}\frac{1+^{i-1}v}{^{i-1}E}{^{i-1}}T + \frac{^{i}E}{1+^{i}v}(\Delta e - \Delta\varepsilon^p - \Delta\varepsilon^c - \Delta\varepsilon^{Trp}) \qquad (5\text{-}31)$$

$$^{i}\sigma_m = \frac{^{i}E}{1-2^{i}v}(^{i}\varepsilon_m - {^{i}}\varepsilon^{th} - {^{i}}\varepsilon^{Trp})$$

Step 4: Rotate the Cauchy stresses to the current orientation; and finally a translation to point i.

$$^{i}\sigma = {_{i-1}^{i}}R^{i-1}(^{i}\sigma)_{i-1}^{i}R^{T} \qquad\qquad (5\text{-}32)$$

Numerical Results

The current formulation has been tested in several ways. Thermal expansion and contraction were tested within the elastic range. The mesh size was *15.8 x 3.2 x 3 mm*, divided evenly along the *x*-direction into 5 elements. Some results are shown in Figure 5-10.

The mid-element, element *3*, had a temperature increment of *30°C*. Although either a Lagrangian formulation or a steady state formulation reaches the same results in this case, the results have a different meaning. All elements but element 3 were kept at constant room temperature in the Lagrangian formulation. While a thermal expansion occurred in element *3*, there was no thermal expansion or shrinkage in any other elements. In the steady state formulation, the material in element *4* and element *5* had experienced a complete thermal cycle, heated up to *50°C* and then cooled down to *20°C*. A corresponding cycle of expansion and contraction occurred in these two elements. This example tested the elastic recovery of the model.

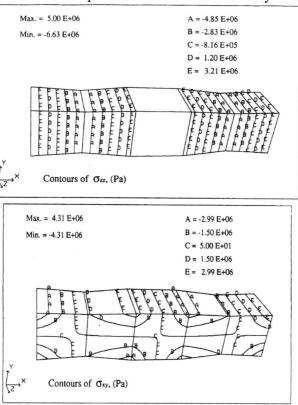

Figure 5-10: Thermal expansion of a middle element in a 5 element bar. The two ends are fixed in *x* direction. The deformation scale factor is 50.

A full scale computation was performed to analyze an edge weld on a thin flat bar. This example had been analyzed with the Lagrangian formulation [27]. The mesh in Figure 5-11 is created on a piece of material cut from the long strip, which contained a steady state temperature field. The mesh size is *543 x 50.8 x 6.35 mm.*

In the transient analysis, this temperature distribution is shown at *100.5* seconds after welding started. With medium resolution in space and time, *65* steps in the Lagrangian formulation were required to reach this state for temperatures and stresses.

The material used is low carbon steel. The temperature dependent mechanical and thermal properties of *AISI 1020* are used. Weld speed is set at *5 mm/s*. The heat source is a prescribed temperature field in a weld pool with a double ellipsoid shape. The source had length *8 mm*, width *5 mm*, and depth *3 mm*. The resulting computed power input is *0.32 KJ/mm*. The heat efficiency is estimated to be *70* percent.

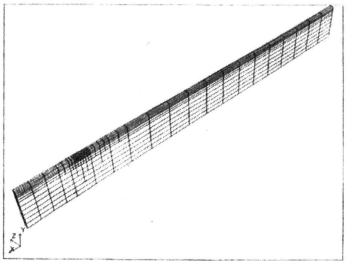

Figure 5-11: FEM mesh designed for simulating the edge weld in the steady state, from [6 and 25].

Figure 5-12 shows the temperature field on the top and bottom faces of the plate.

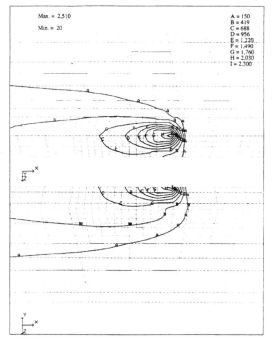

Figure 5-12: Contours of temperature °C around the weld pool area. Top: symmetry plane; Bottom: outer surface.

As often seen in a welding shop, thermal distortion bends the bar into the "S" shape seen in Figure 5-13. The displacements in this Figure have been magnified *30* times. In a welded structure, the residual stresses are highly localized, as is the element distortion around the weld path. Elements far from the weld path deform largely by rigid body rotations.

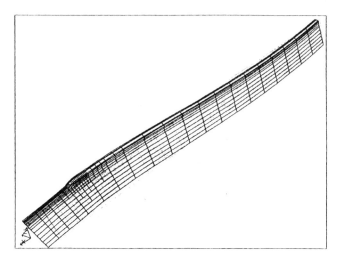

Figure 5-13: Distortion of the mesh after an edge weld. All displacements are enlarged *30* times. The undeformed mesh before welding is shown in grey lines.

A detailed picture of the σ_{xx} stress around the weld pool is given in Figure 5-14. For this weld the longitudinal stress σ_{xx} is the most significant one, with an average value 3 times larger than the other five stress components. It is the longitudinal stress that caused most of the distortion shown in Figure 5-14.

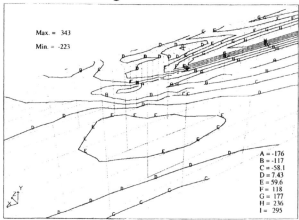

Figure 5-14: Contours of stress, σ_{xx} (MPa), round the weld pool area. The phase transformation causes negative stress inside D after the weld pool.

Figure 5-15 shows the computed values of the through-thickness average longitudinal stresses, comparing to the measured stresses and the stresses computed in the Lagrangian formulation. The measured stresses and the stresses computed in the Lagrangian formulation were provided by Oddy et al. [27].

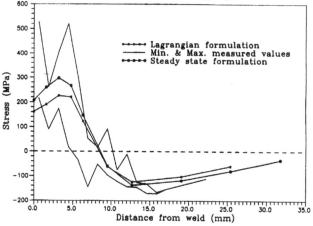

Figure 5-15: Through-thickness average of longitudinal stress σ_{xx} versus distance from welded edge. For the Lagrangian formulation [15], sampling data is from the mid-length of the weld path. For the current formulation, the data is from the near end of the path.

The computational results fit in the range of the measured data quite well. The steady state formulation has the better results near and far away from the weld center line. This difference is actually a problem of computational costs. In order to reduce computer time, the convergence tolerance was set too large in the analysis of reference [27] .With a more powerful computer and a smaller convergence tolerance, a more recent analysis in the Lagranian formulation, performed by Oddy et al [27], has shown similar results to those obtained in this steady state analysis.

The cost of computing the steady state solution is approximately the same as the cost of one transient Lagrangian step. If we use the Lagrangian method to compute the starting and stopping transient, but use the Eulerian method to compute the steady state part of the weld, the Eulerian method is more than 2000 times cheaper for the steady state portion of the 10 m weld. It is more than 2000 times cheaper because it need not remesh for each time step. Further more the mesh can be designed for the weld pool shape. More important, filler metal can be added naturally in this formulation. Figures 5-13 and 5-16 show the results of this analysis.

Figure 5-16: Steady state analysis of a butt weld with filler metal addition. The undeformed mesh is shown in faint dotted lines. The transient temperature for this analysis was shown in Figure 3-21.

In most thermal analyses of welds performed to date, the mesh is not fine enough to resolve the rapid rise in temperature in front of the weld pool. The effect is that the computed strain rate can be much less than the real strain rate. Consequently the computed stress can be much less than the real stress.

When the convergence rate is described by the number of iterations of a nonlinear solver at a fixed tolerance, the test cases demonstrated that the convergence rate of the steady state formulation was the same as the convergence rate of the Lagrangian formulation for one time step. With this convergence rate the cost of a steady state analysis is the same as the cost of one time step in the transient analysis.

5.2.3 Linear Viscous Isotropic Plasticity around the melting point

The main difference in implementing a linear viscous or any other kind of rate dependent model compared to a rate independent model arises in the evolution equation for stress and deformation resistance.

As we stated before the linear viscous model is a particular instance of the plastic potential function (5-4) with rate sensitivity $m=1$:

$$\Phi = \dot{\varepsilon}_0 \{\frac{\tilde{\sigma}}{s}\}^2 \tag{5-33}$$

In the linear viscous model, the plastic strain rate is:

$$\tilde{\dot{\varepsilon}}^P = (1/v)\tilde{\sigma} \tag{5-34}$$

where $v = \{s/\dot{\varepsilon}_0\}$ is the viscosity in a linear viscous Maxwell body [37].

Then the relaxation equation yields:

$$\tilde{\sigma} + (\frac{v}{3\mu})\dot{\tilde{\sigma}} = 0 \tag{5-35}$$

Taking the trial effective stress $\tilde{\sigma}^*$ as initial condition and integrating (5-34) over the time step we get:

$$\tilde{\sigma} = \tilde{\sigma}^* \exp(-\frac{\Delta t}{r^*}) \tag{5-36}$$

where:

$$r^* = v/3\mu \tag{5-37}$$

is the relaxation time.

We estimate the macroscopic viscosity v for deformation of polycrystals at high temperatures to be:

$$v = \frac{kT}{v_D \rho |b| \zeta v^*} \exp(\frac{\Delta G}{RT}) \tag{5-38}$$

The pre-exponential factor in (5-38) and the activation energy ΔG are data to be found experimentally. The most important data is ΔG that determines the order of magnitude of υ. For plasticity above *0.8* T_m, where T_m is the melting point in Kelvin degree, it was found that for relatively coarse grain structures the activation energy for self diffusion is a good approximation for ΔG. This suggests that the effective mechanism of dislocation slip is diffusion controlled. Substituting default values for the rest of the unknowns does not affect the right order of magnitude of υ. Reasonable values for those defaults are the microscopic activation volume $v^* = 2|b|^3$, Debeye frequency of thermal fluctuation $v_D = 10^{12}$ *(1/s)*, the density of mobile dislocations $\rho = 10^{12}$ (1/m²), the average jump distance of a dislocation segment in one activation event $\zeta = |b|$ and the Burger's vector length $|b| = a\sqrt{2}/2$ for *FCC* and $|b| = a\sqrt{3}/2$ for *BCC* lattice, respectively, where a is the crystal lattice parameter.

5.2.4 Rate dependent isotropic plasticity (General case)

For steel at temperatures from *700* to *1300* ℃ we adopt a rate dependent plasticity model using the constitutive functions proposed by Brown [38] for the effective plastic strain rate was adopted in [31]:

$$\dot{\tilde{\varepsilon}}^p = A\exp(-\frac{\Delta G}{RT})\left[\sinh(\xi\frac{\tilde{\sigma}}{s})\right]^{(1/m)} \tag{5-39}$$

and for evolution of the internal variable:

$$\dot{s} = \{h_0|(1-\frac{s}{s^*})|^l \, sign(1-\frac{s}{s^*})\}\dot{\tilde{\varepsilon}}^p \tag{5-40}$$

where the temperature dependent term in (5-39), A exp(-ΔG/RT), is the reference (for a given temperature) strain rate, m is a rate sensitivity, the hyperbolic sine accounts for the contribution of stress to thermal activated slip, h_0 is the reference hardening parameter and s^* is the saturation value of deformation resistance s:

$$s^* = \tilde{s}\left[\frac{\dot{\bar{\varepsilon}}^p}{A}\exp(\frac{\Delta G}{RT})\right]^n \tag{5-41}$$

Note that A , h_0 , $\tilde{s}$ exponents m and l and dimensionless activation volume ξ, generally, are temperature and microstructure dependent material properties.

5.2.5 Changing Constitutive Equations in Time and in Space

As stress evolves with time, a new problem is solved for each time step. Suppose that in time step n, rate dependent plasticity is used at Gauss point m. Then in time step $n+1$, suppose rate independent plasticity is used at Gauss point m. This discontinuity or switch of the constitutive equations does not cause any problems. The reason is that the initial conditions required for each time step are the initial geometry, initial stress, initial strain, the boundary conditions from t_n to $t_n + dt$ and the constitutive equations in the interior of the time step. If the time step is from t_n to $t_n + dt$, the constitutive equation must be defined only for time t such as $t_n < t <= t_n + dt$. There is no need to define the constitutive equation at times earlier than t_n.

Suppose a rate independent constitutive equation is used at one Gauss point and a rate dependent constitutive equation is used at a neighboring Gauss point. This does not cause any problems in the analysis. At each Gauss point, the analysis gives the Gauss point an initial stress, values of the internal variable such as yield stress or deformation resistance and strain rate and asks for the stress at that point at the end of the time step. Each Gauss point can have its own constitutive equations and it causes no problems and introduces no complexity into the analysis other than providing the capability of switching between constitutive equations and providing the mappings at internal variables required by each constitutive equation such as yield strength to deformation resistance.

5.2.6 Numerical Experiments and Results

Figure 5-17 shows the behavior of four constitutive models. Their properties are shown in Tables 5-2 and 5-3. Two models use rate independent plasticity, one uses rate dependent plasticity and one uses linear viscous plasticity. The test is uniaxial and the strain rate is 0.005 s^{-1} in the time interval [0, 4] seconds. The strain rate is zero in the time interval [4, 10]. The strain rate is -0,005 s^{-1} in the time interval [10, 13]. The strain rate is zero in the time interval [13, 25]. Note that in the rate independent models there is no stress relaxation. In the rate dependent model the stress relaxes toward the deformation resistance. In the linear viscous model, the stress decays toward zero. The viscosity used in Figure 5-17 is larger than that used in the welding test. If the viscosity was not increased, the stress would appear to relax instantly in the time scale used in Figure 5-17.

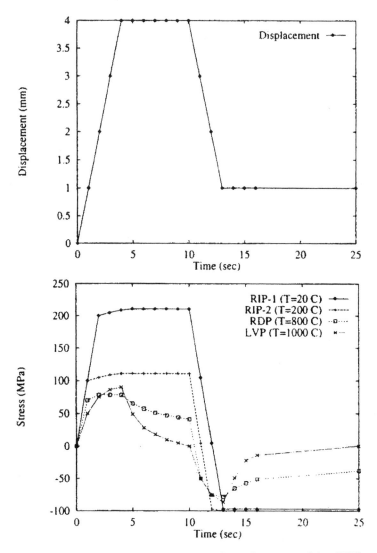

Figure 5-17: The response of two rate independent materials (RIP), a rate dependent material (RDP) and a linear viscous material (LVP) to sequence of strain rates is shown [36].

Table 5-2: Material properties for rate independent plasticity (RIP) and linear viscous plasticity (LVP)

	RIP-1	RIP-2	LVP
Young's Modulus (GPa)	200	200	200
Poisson Ratio	0.3	0.3	0.3
Hardening Modulus (GPa)	2	2	0
Yield Stress (MPa)	200	100	--
Viscosity (GPa.s)	--	--	200

Table 5-3: Material properties for rate dependent plasticity (RDP)

Material Parameter	Value	Material Parameter	Value
$A(s^{-1})$	6.346×10^{11}	ξ	3.25
ΔG (J/mole)	3.1235×10^5	m	0.1956
$\widetilde{s}$ (Pa)	1.251×10^8	n	0.06869
h_0 (Pa)	3.0931×10^9	1	1.5
a (m)	3.50×10^{-10}		

In the weld analyses a prescribed temperature weld heat source was used [43] to model the arc.

Three and ten seconds after starting the weld Figures 5-18, 5-19, 5-20 and 5-21 show the temperature distribution, sub-domains for each constitutive equation, effective plastic strain, pressure and several stress components. Plate dimensions *5×5×0.2 cm*. Weld pool dimensions *1.5 (long) ×1.5 (wide) ×0.5 (deep) cm*. Welding speed *6 ipm (0.25 cm/s)*. Maximum temperature was *2000 °C* and solidus temperature was *1500°C*. A *15×15×1* element mesh was used. Although the mesh is relatively coarse, it does demonstrate the expected behavior. The high compressive ridge in front of the weld is in the rate independent plasticity sub-domain. It is possible that the yield strength used should be corrected for this strain rate. The longitudinal strain pulling material back from the weld pool is another interesting feature.

Other cases were also analyzed. They differed only in weld speed. The slow weld speed was *6 ipm* and the fast speed was *20 ipm (0.25 and 0.83 cm/s)*, Figure 5-22. In the *20 ipm* weld the mesh is not fine enough to resolve the thermal shock and the constitutive equations change directly from rate independent to linear viscous. The longitudinal strain pulling material back from the weld pool is an interesting feature.

Due to a slight change in mesh distribution on the left and right side of the weld (element row *8* or *7*) the temperature field on the left and right side is not symmetric. It is shown that it causes no problems in the analysis by switching between constitutive equations.

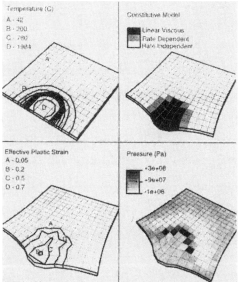

Figure 5-18: At three seconds after starting the weld, the transient temperature, constitutive equation type, effective plastic strain and pressure are shown.

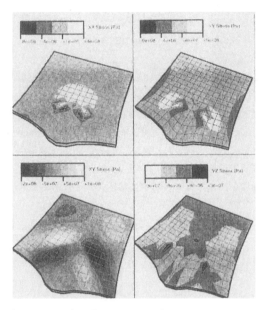

Figure 5-19: At three seconds after starting the weld, $\sigma_{xx}, \sigma_{yy}, \sigma_{xy}$ and σ_{yz} are shown.

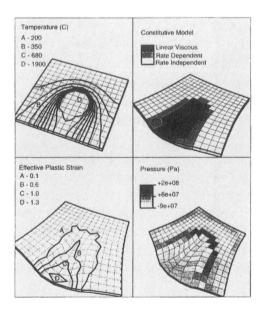

Figure 5-20: At ten seconds after starting the weld, the transient temperature, constitutive equation type, effective plastic strain and pressure are shown.

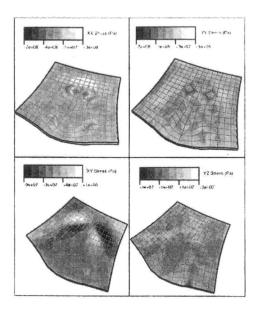

Figure 5-21: At ten seconds after starting the weld, $\sigma_{xx}, \sigma_{yy}, \sigma_{xy}$ and σ_{yz} are shown.

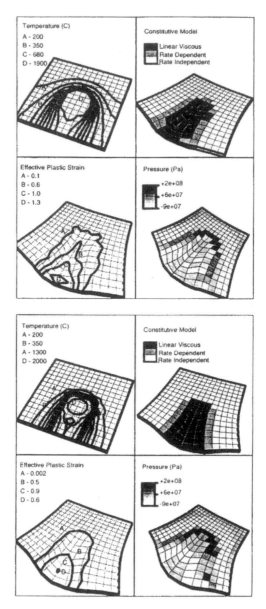

Figure 5-22: At ten seconds after starting the weld, the transient temperature, constitutive equation type, effective plastic strain and pressure are shown for the *6* and *20 ipm* welding speeds.

The weld direction is the *y*-direction, the upward normal on the plate is the *z*-direction and the *x*-direction is transverse to the weld. Figure 5-23 shows the displacements transverse to the weld at various distances from one node, *(1/3 cm)* from the weld centerline extends from nodes *5* to *9* in the slow weld and from nodes *3* to *15* in the fast weld.

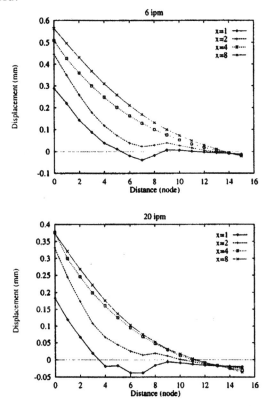

Figure 5-23: The abscissas are nodes numbered from left to right. The distance between nodes is *1/3 cm*. When the weld arc reaches the center of the plate, the weld pool is centered at node *8* and extends from node *5.5* to node *10.5* on the plate. At this instant the transverse displacements along lines parallel to the weld at distances of *1, 2, 4* and *8* nodes from the weld centerline are shown. The welding speeds of *6* and *20 ipm* are shown above each plot.

Figure 5-24 shows the transverse stress, σ_{xx}, to the weld on the weld centerline. Chihoski was not able to measure stress. Data of this type would be also useful to compare with Sigmajig test data.

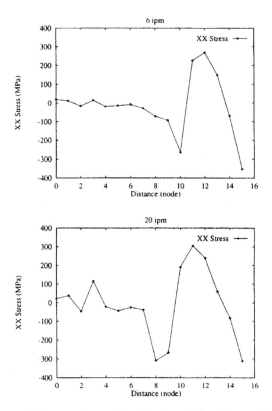

Figure 5-24: The welding speeds are *6* and *20 ipm*. When the weld arc reaches the center of the plate, the transverse stress σ_{xx} is shown along the weld line

Also the compressive stress region in Figure 5-24 is larger for the faster weld. This is the effect observed by Chihoski and suggests the faster weld would be less susceptible to hot cracking because more of the region susceptible to hot cracking is in compression. The qualitative agreement is as good as could be expected because the welds are quite different.

This model appears to be useful step towards quantitative analysis of stress and strain in the region susceptible to hot cracking in welds. A more realistic quantitative analysis of hot cracking will require a finer mesh to resolve the thermal shock in front of the weld pool. A larger plate is needed to provide more constraint to the weld pool region.

We next consider one or more algorithms for computing flow lines or streamlines associated with a steady state weld.

References

1. Lindgren L-E. Finite element modeling and simulation of welding Part I Increased complexity, J of Thermal Stresses 24, pp 141-192, 2001
2. Goldak J. A. Modeling thermal stresses and distortions in welds, Recent Trends in Welding Research, Ed. David S.A. and Vitek J, (Materials Park, OH, ASM), pp 71-82, 1990
3. McDill JMJ, Oddy AS, Goldak JA and Bennisson S. Finite element analysis of weld distortion in carbon and stainless steels, Journal of Strain Analysis for Engineering Design, Vol. 25, No. 1, pp 51-53, 1990
4. Ueda Y. and Yamakawa T. Analysis of thermal elastic-plastic stress and strain during welding by finite element method, JWRI, Vol. 2, No. 2, pp 90-100, 1971
5. Goldak J.A., Oddy A., Gu M., Ma W., Mashaie A and Hughes E. Coupling heat transfer, microstructure evolution and thermal stress analysis in weld mechanics, IUTAM Symposium, Mechanical Effects of Welding, June 10-14, Lulea Sweden, 1991
6. Gu M. and Goldak J.A. Steady state formulation for stress and distortion of welds, J of Eng. For Industry, Vol. 116, pp 467-474, Nov. 1994
7. Goldak J.A., Breiguine V., Dai N., Hughes E. and Zhou J. Thermal Stress Analysis in Solids Near the Liquid Region in Welds. Mathematical Modeling of Weld Phenomena, 3 Ed. By Cerjak H., The Institute of Materials, pp 543-570, 1997
8. Hirt C. W., Amsden A. A. and Cook J. L. An arbitrary Lan- grangian-Eulerian computing method for all flow speeds," Journal Computational Phys., Vol. 14, No.3, pp 227-253, 1974
9. Donea J. Arbitrary Lagrangian-Eulerian finite element methods, computational methods for transient analysis, T. Belystschko T. and Hughes T.J.R., eds., Computational Methods in Mechanics, Vol. 1, Elsevier Science Publishers B. V., pp 473-516, 1983
10. Haber R. B. A mixed Eulerian-Lagrangian displacement model for urge-deformation analysis in solid mechanics, Computer Methods in Applied Mechanics and Engineering, Vol. 43, pp 277-292, 1984
11. Koh H. M. and Haber R. B. "Elasto dynamic formulation of the Eulerian-Lagrangian kinematic description," ASME Journal of Applied Mechanics, Vol. 53, No.12, pp 839-845, 1986
12. Ghosh, S. and Kikuchi N. An arbitrary Lagrangian-Eulerian finite element for urge deformation analysis of elastic-visco-plastic solids," Computer Methods in Applied Mechanics and Engineering, Vol. 86, pp 127- 188, 1991

196   *Computational Welding Mechanics*

13. Paul R. and Dawson: On modeling of mechanical property changes during flat rolling of aluminum, Int. in Solids Structures, Vol. 23, No.7, pp 947-968, 1987

14. Bergheau J. M., Pont D. and Leblond J. B. Three-dimensional simulation of a laser surface treatment through steady state computation in the heat source's commoving frame, Mechanical Effects of Welding, IUTAM Symposium, Lulea Sweden, June 1991

15. Kojic M. and Bathe K.-J. The Effective-Stress-Function algorithm for thermo-elastic-plasticity and creep, Int. Num. Math. Engineering, Vol. 24, pp 1509-1532, 1987.

16. Bathe, K. J., Kojic M. and Walczak J. Some developments in methods for large strain elasto-plastic analysis, computational plasticity models software and applications I, D. R. J., Owen et al., eds., Pineridge Press, Swansea U. K., pp 263-279.

17. Oddy AS, Goldak JA and McDill JMJ. Transformation effects in the 3D finite element analysis of welds," in David S. A. and Viterk J.M., eds., Recent Trends in Welding Science and Technology TWR '89, ASM International, Materials Park OH, pp 97-101,1989

18. Hughes T.J.R. Numerical implementation of constitutive models: rate independent deviatoric plasticity, Proceedings of the Workshop on the Theoretical Foundation for Large-Scale Computations of Nonlinear Material Behavior, Northwestern Univ. Evanston IL, Oct. 24, 1983

19. Belytschko T. and Hughes T. Computational analysis for transient analysis, Vol. I, Computational Methods in Mechanics, North-Holland New York USA, pp 22-37, 1983.

20. Dienes J. K. On the analysis of rotation and stress rate in deforming bodies, ACTA Mechanic, Vol. 32, pp 217-232, 1979.

21. Hill R. Aspects of invariance in solids mechanics, Advances in Applied Mechanics, Vol. 18, Academic Press, New York USA, pp 1-75, 1978.

22. Malvern L.E. Introduction in the mechanics of a continuous medium, Prentice-Hall Inc., Englewood Cliffs, New Jersey, 1969

23. Gurtin, M. E., An introduction in continuum mechanics, Academic Press, New York.

24. Hoger, A., Carlson, D.E. Determination of the stretch and rotation in the polar decomposition of the deformation gradient, Quart. of Appl. Math., Brown University, pp 113-117, April 1984

25. Gu M. (1992). Computational weld analysis for long welds. Doctoral thesis Carleton University.

26. Masubuchi, K.Analysis of welded structures, N. Y., Pergamon Press, pp 119-120, 1980

27. Oddy, A.S, McDill, JMJ., and Goldak, JA. Consistent strain fields in 3D finite element analysis of welds, ASME Journal of Pressure Vessel Technology, August, pp 309-311, August 1990

28. Goldak J.A. and Gu M. Computational weld mechanics of the steady state, Mathematical Modeling of Weld Phenomena 2, Ed. H. Cerjak, The Institute of Metals, pp 207-225, 1995

29. Gu M.,Goldak J.A. Steady state thermal analysis of welds with filler metal addition, Can. Met., Vol. 32, pp 49-55,1993

30. Matsunawa A. Modeling of heat and fluid flow in arc welding, Recent Trends in Welding Science and Technology, Ed. David S.A. and Vitek J (Materials Park, OH, ASM), pp 1-12, 1993

31. Goldak J.A., Zhou J., Breiguine V. and Montoya F. Thermal stress analysis of welds from melting point to room temperature, JWRI, Vol. 25, No. 2, pp 185-189, 1996

32. Goldak J.A., Breiguine V. and Dai N. Computational weld mechanics; A progress report on ten grand challenges, International Trends in Welding Research, Gatlinburg Tennessee, June 5-9 1995

33. Weber G. and Anand L. Finite deformation constitutive equations and a time integration procedure for isotropic, hyperelastic-viscolplastic solids. Comput. Methods Appl. Mech. Engrg. 79, pp 173-202, 1990

34. Richter F: Die wichtigsten physikalischen Eigenschaften von 52 Eisenwerkstoffen, Heft 8, Stahleisen Sonderberichte, Verlag Stahleisen, Duesseldorf Germany, 1973

35. Weber G., Lush A., Zavaliangos A. and Anand L. An objective time-integration procedure for isotropic rate independent and rate dependent elastic-plastic constitutive equations, International Journal of Plasticity 6, pp 701-774, 1990

36. Goldak J.A., Breiguine V., Dai N. and Zhou J. Thermal stress analysis in welds for hot cracking, ASME Journal of Pressure Vessel Technology, Jan. 24, 1996

37. Le Tallec P. Numerical analysis of viscoelastic problems, Masson. Paris, 1990

38. Brown S.B., Kim K.H. and Anand L. An internal variable constitutive model for hot working of metals, International Journal of Plasticity, Vol. 5, pp 95-130, 1989

39. Oddy AS, Goldak JA and McDill JMJ. Transformation plasticity and residual stresses in single-pass repair welds, ASME J Pressure Vessel Technology, Vol. 114, pp 33-38, 1992

40. Oddy AS, Goldak JA and McDill JMJ. Numerical analysis of transformation plasticity relation in 3D finite element analysis of welds, European Journal of Mechanics, A/Solids, Vol. 9, No. 3 pp 253-263, 1990

41. Leblond J.B., Mottet G. and Devaux J.C. A theoretical and numerical approach to the plastic behavior of steels during phase transformation –II. Study of classical plasticity for ideal-plastic phase, Journal Mech. Phys. Solids, Vol. 34, pp 411-432, 1986

42. Oddy AS, Goldak JA and Reed RC. Martensite formation, transformation plasticity and stress in high strength steel welds, Proc. of the 3rd Int. Conf. on Trends in Welding Research, pp 131-137, 1992

43. Goldak J.A. Bibby M.J. and Gu M. Heat and fluid flow in welds, Proceedings of the International Institute of Welding Congress on Joining Research, Ed. T.H. North, Chapman and Hall, pp 69-82, July 1990

44. Ashby M.F. Physical modeling of materials problems, Materials Science and Technology, Vol. 8, pp 102-111, 1992

45. Radaj D. Eigenspannungen und Verzug beim Schweissen, Rechen- und Messverfahren, Fachbuchreihe Schweisstechnik, DVS-Verlag GmbH, Duesseldorf 2000

46. Matsuda F., Tomita S. Quantitative evaluation of solidification brittleness of weld metal by MISO technique, Recent Trends in Welding Science and Technology , Ed. David S.A and Vitek I., (Materials Park, OH, ASM), pp 689- 694, 1993

47. Chihoski Russel A. Expansion and Stress Around Aluminum Weld Puddles, Welding Research Supplement, pp 263s-276s, Sep. 1979

48. Pilipenko A (2001) Computer simulation of residual stress and distortion of thick plates in multielectrode submerged arc welding. Doctoral thesis, Norwegian University of Science and Technology

49. Hibbitt H.D., Marcal P.V. A numerical thermo-mechanical model for the welding and subsequent loading of a fabricated structure, Comp. & Struct., Vol. 3, pp 1145-1174, 1973.

50. Gu M., Goldak J.A. and Hughes E. Modeling the evolution of microstructure in the heat-affected-zone of steady state welds, Can. Metall. Quarterly 32, No. 4, pp 351-361, 1993

51. Patel B. (1985). Thermo-elasto-plastic finite element formulation for deformation and residual stresses due to welds. PhD Thesis, Carleton University.

Chapter VI

Carburized and Hydrogen Diffusion Analysis

6.1 Introduction and Synopsis

It is widely known that hydrogen is extremely harmful to the safety of a metal construction. It influences the mechanical properties of base metal or a joint or, directly results in fracture during welding due to cracking, so called hydrogen induced cracking *HIC* or hydrogen assisted cracking *HAC*. While the time to cool from *800 °C* and *500 °C* for steels determines the hardening, the time to cool to *100 °C* determines hydrogen retained in the metal. Slower cooling times allow more hydrogen to escape to the air.

HAZ cold cracking, sometimes called underbead cracking, Figure 6-1, is a well known phenomenon in welding technology.

Figure 6-1: Hydrogen underbead (assisted) cracking in the HAZ of a single-pass, shielded-metal arc bead-on plate weld deposited on a C-Mn microalloy steel, from [28]

Cracks are caused essentially by residual tensile stresses that form as a result of thermal shrinkage when a weld cools. The residual stress level often exceeds the yield strength of the material which in turn causes plastic flow. When the plastic capacity of the joint region is exhausted cracks can form. The situation is complicated by the simultaneous formation of hardened microstructures of limited plastic capacity. Moreover, it is further complicated by the presence of hydrogen. It has long been recognized that necessary though not sufficient conditions for *HIC* are:
   a) Sufficiently high hydrogen concentration,
   b) Sufficiently sensitive microstructure,
   c) Sufficiently high tensile stress.
However, there is disagreement on the values of numbers to define sufficiently.

Hydrogen forms in the arc and is "pumped" into the *HAZ* through the weld metal. In this model the effect of hydrogen is assumed to dominate the mechanism. Hydrogen diffuses into the *HAZ* at high temperatures but escapes as the weld cools.

The hypothesis is that a critical level of hydrogen exists in the *HAZ* below which cold cracks do not form. This critical level depends on composition, joint shape, workpiece size and shape, yield level of the base material and ambient temperature.

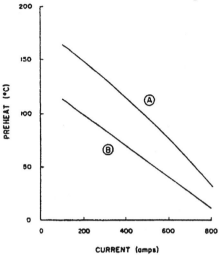

Figure 6-2: Preheat to avoid cracking computations; submerged arc welding: voltage *25V*, welding speed 5mm/s, ambient temperature *10°C;* local preheat conditions: heater width *100mm*, heater strength *0.1W/mm,* from [22 and 23].

In practice preheat is used to slow down the cooling and provide an opportunity for the *HAZ* hydrogen to drop below the critical level [22 and 28]. At temperatures below roughly 100 to 200 °C, the diffusion of hydrogen can be dominated by diffusion along the gradient of the hydrostatic stress toward high hydrostatic (tensile) regions. Below this temperature diffusion is so slow that any hydrogen left can be considered to be trapped. At the same time cracks form at temperatures below *100 °C* and therefore the hydrogen environment is fixed before cracking occurs. A critical time is defined in the Yurioka model [22] which is the cooling time to *100°C*, i.e., the time for hydrogen to drop to the critical level. If the real cooling time is greater than the critical cooling time then cracking is avoided.

The real cooling time of the weld can be calculated with the help of a long time low temperature heat flow model. Such computations were used to generate the data presented in Figure 6-2. This Figure is submitted to show just how useful the system can be in practice.

The preheat level needed to avoid underbead cracking for (A) uniform preheat (B) preheat with electrical strip heaters is shown.

The cracking mechanism is very complex and it will be some time before it is well enough understood to model it precisely. Nevertheless this model is an important development and the computations form a useful basis for engineering judgment in practice.

Pipeline designers are aware that hydrogen induced cracking can cause failures and they are anxious to specify designs that minimize the risk of failure. Late in the life of a pipeline, *HIC* is often associated with corrosion. Early in the life of a pipeline or a repair, *HIC* is often associated with welds. Specifically, our objective is to develop criteria for the design of weld procedures that minimize the risk of failure due to *HIC*.

It is desirable to weld natural gas pipelines under pressure both for making new connections to the pipeline, called hot-taps, and to repair damage to the pipeline. Welding on the pressurized pipeline avoids the costs of shutting down the pipeline and depressurizing. In thinner walled pipes, there is a risk of burn-through when welding while the pipeline is pressurized. Should burn-through occur, the safety of the welding personnel are at risk and the pipe will have to be shut down to repair the burn-through.

Kiefner [1 and 2] studied the problem in some detail. He recognized that the risk of burn-through was greatest in thin wall pipes. He also proposed a model to predict the risk of burn-through. The model used a rather simple *2D* finite difference method with forward Euler time integration for the thermal analysis. If the maximum internal wall temperature exceeded 982 C (*1800 °F*), then the risk of burn-through was considered high. Kiefner included the effect to flowing the gas or liquid on the thermal analysis through the convection coefficient. However, he did not consider the effect of stress and deformation.

Displacement, strain and stress are computed by solving the conservation of linear momentum and mass equations. If the pressure and temperature are high enough, the wall thin enough and the time at temperature long enough, a groove can form under the weld by visco-plastic flow, i.e., creep. Such a groove changes the wall thickness and the distance from the bottom of the weld pool to the inner wall of the pipe and the temperature on the inner wall of the pipe. This will change the carburized layer thickness and composition. This coupling or interaction between thermal and stress analysis is an important nonlinear effect that has usually been neglected in computational weld mechanics.

Burn-through is primarily a high temperature creep phenomena and requires modeling viscous flow. Solving the energy equation on the current deformed geometry captures an important nonlinear coupling between stress analysis and energy analysis.

6.2 Carburization; Theory and Numerical Methods

The computational weld mechanics *CWM* analysis of the process of welding on pressurized pipelines involves several difficult problems that make this a particularly challenging problem.

The greatest challenge is to develop the capability to deal with creep at temperatures in the range *900 °C (1652 °F, 1173 °K)* to *1500 °C (2732 °F, 1773 °K)*. For sufficiently short times at temperatures below *900 °C (1652 °F, 1173 °K)*, the creep rate is sufficiently slow that during the rather short times at high temperatures during welding creep can be ignored in most problems and rate independent plasticity is the appropriate material behavior. However, it is exactly at temperatures above *900 °C (1652 °F, 1173 °K)* that internal pressure can cause burn-through. It is exactly in this temperature range that the steel deforms primarily by viscous flow, which is rate dependent plastic or visco-plastic or creep. In this temperature range, a realistic analysis must deal with viscous flow and use realistic values of the viscosity, the material parameter that relates the deviatoric strain-rate to stress.

A second challenge is the need to couple the thermal analysis with the stress analysis more strongly than is usually done in *CWM*. In particular, the thermal analysis must be done on the geometry deformed by the stress analysis.

A third challenge is to deal with the length scales that range from a fraction of a *mm* near the weld pool to a *3.66m (12 ft.)* long pressure vessel length. This large range of length scales can make the *FEM* problem numerically ill-conditioned. Special solvers that were sufficiently robust were needed to deal with this problem.

A fourth challenge is to manage the complexity of the software and the complexity of input data required for analyses. The addition of filler metal is desirable to provide a more complete and realistic simulation.

While welding on the outer wall of a pressurized natural gas pipeline, the heat from the weld can cause localized carburization on the inner wall of steel pipe. The carbon levels can exceed the eutectic composition of approximately *4.3* weight percent carbon in iron. This phenomenon can produce a thin layer of liquid cast iron on the internal wall of the pipe directly under the welding arc.

The diffusion of carbon is assumed to depend primarily on the spatial gradient of carbon composition and the temperature through the temperature dependence of the diffusivity of carbon. Local equilibrium is assumed at the liquid-solid interface.

Natural gas pipelines are welded under pressure for two reasons; to make a new connection to an existing pipeline or to repair damage such as that caused by corrosion [1]. If the pipeline is sufficiently thin, the part of the inner wall under the arc can reach temperatures approaching the melting point, *1530 °C* for a typical *HSLA* pipeline steel. At these temperatures natural gas, which is mostly methane CH_4, can decompose by the reaction $CH_4 \rightarrow C + 2H_2$. The resulting carbon atoms react with the surface of the pipe. We conjecture that at temperatures above the eutectic temperature for a typical *HSLA* pipeline steel, *1147 °C*, a very thin *Fe-C* film forms, perhaps only *1* to *2* atoms thick at first, and then the film grows by diffusion of carbon. The carbon diffuses into this film from the gas atmosphere, through the liquid film across the liquid-solid interface and into the solid. In this very thin film of liquid, it is assumed that advection, i.e., stirring or fluid flow effects, can be neglected. A film of such a brittle material increases the risk of cracking in the pipe. The addition of hydrogen from the decomposition of natural gas adds to the concern.

In this chapter, this phenomenon is simulated by the following method. The *3D* transient energy equation is solved to compute the transient temperature field. While the temperature at a point on the internal pipe wall exceeds the eutectic temperature, it is assumed a thin layer of liquid exists. Further it is assumed this layer grows by diffusion of carbon into the underlying solid. The limiting process is assumed to be the temperature dependent diffusion of carbon. A thermal-stress analysis computes distortion which can change the wall thickness and thus the temperature computed in the thermal analysis and the carburization analysis. The addition of hydrogen from the decomposition of natural gas adds to the concern. Modeling the hydrogen behavior will be presented in section 6-3.

The temperature curves obtained at the points on the internal surface of the pipe were used to simulate the growth of the carburized layer. The carburized layer on the internal surface of the pipe was assumed to result from the absorption of carbon by the pipe material from the hydrocarbon rich environment inside the pipe. The increase in the carbon concentration results in melting starting at the internal surface of the pipe at temperatures ($\approx 1147°C$) significantly lower than solidus temperature ($\approx 1500°C$) corresponding to the nominal concentration of carbon in steel, $c_0 \approx 0.1\%$. They also assumed that the rate of carbon absorption is controlled by the diffusion of carbon through the growing liquid layer and that concentrations at the liquid-solid interface, $c_{\gamma L}$ and $c_{L\gamma}$, and liquid-gas phase interface, c_{L0}, are determined by the equilibrium liquidus and solidus concentrations of $Fe - Fe_3C$ diagram, Figure 6-3. Therefore they neglected any effects of the melting kinetics and kinetics of the reaction between the gas phase and liquid material on the growth rate.

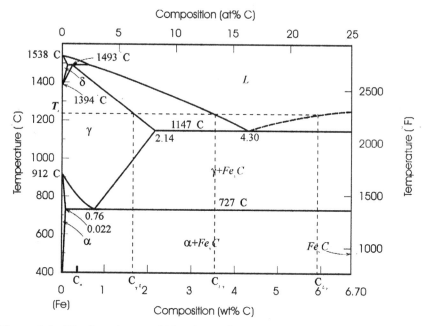

Figure 6-3: The iron–iron carbide phase diagram with interface concentrations shown for liquid layer growing at temperature T_i

It is possible to neglect the absorption of carbon by a solid material at a temperature below the eutectic temperature because the diffusivity of carbon in the solid phase is more than two orders of magnitude lower than the diffusivity of carbon in the liquid. As a result the overall contribution of gas to the solid phase transport into carburization is negligibly small compared to the contribution from the fluxing effect. In this model of the fluxing process the changes in the concentrations and phase state are simulated only at temperatures above the eutectic temperature. It is assumed that an infinitesimally thin layer of the liquid phase is formed at the pipe internal surface at the moment when its temperature becomes higher than the eutectic temperature. After that the layer grows with the rate limited by the diffusion of carbon from the liquid gas interface to the liquid solid interface. The boundary conditions for a corresponding diffusion problem and typical concentration distribution in the system during this process are illustrated in Figure 6-4.

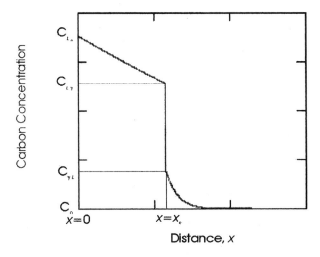

Figure 6-4: The distribution of the carbon concentration in the system with a liquid layer growing from the surface *(x=0)*. The liquid–solid interface is located at $x = x_s$.

An analytical solution exists for the diffusion problem with a constant growth temperature (constant values of $c_{L0}, c_{L\gamma}$ and $c_{\gamma L}$), [4]. Under these conditions the thickness of the growing layer at time t can be found as:

$$x_s(t) \equiv 2\beta\sqrt{D_l t} \tag{6-1}$$

where the coefficient β is determined by the transcendental equation:

$$(c_{\gamma L} - c_{L\gamma})\sqrt{\pi}\beta = (c_{L\gamma} - c_{L0})\frac{\exp(-\beta^2)}{erf(\beta)} + (c_{\gamma L} - c_0)\frac{\sqrt{D_s}}{\sqrt{D_l}}\frac{\exp(\frac{-\beta D_l}{D_s})}{erf(\beta\frac{D_l}{D_s})} \tag{6-2}$$

where D_l and D_0 are diffusion coefficients of the liquid and solid phases, respectively ($D_l = 2.0\times10^{-8} m/s^2$ and $D_0 = 1.0\times10^{-9} m/s^2$ [5]). Equations (6-1) and (6-2) are used to estimate the total thickness of the liquid layer which can be obtained under different welding conditions and to determine the size of mesh cells and number of cells in a mesh for the numerical solution of the diffusion problem with a varying growth temperature.

The numerical algorithm that has been used for the solution of the diffusion problem is based on the cellular model of the solidification process [6 and 7]. In this model space is divided into cells and every cell is characterized by a fraction of liquid, f_l and fraction of solid, $1-f_l$, phases and carbon concentration in the liquid, c_l and/or in solid c_s, phase. The size, Δx, and the number of cells in the *1-D* mesh were estimated using (6-1) and (6-2) so that the liquid layer at the end of the growth process would consist of at least *10* cells and the total length of the mesh would be *2-3* times larger than the final thickness of the liquid layer. The diffusion transport is described in this model by an explicit finite difference scheme implemented separately for liquid and solid parts of the system. The solution for the concentration in liquid is evaluated only in the cells with a non-zero fraction of liquid, and the solution for the concentration in solid is evaluated in the rest of the system. The liquid and solid concentration fields overlap in one cell in which melting occurs and the fraction of liquid gradually increases. The length of time step, Δt, was limited by the stability criterion and by the solidification limit allowing not more than *0.1* of a cell to be solidified in one step.

Different boundary conditions were implemented at the external boundaries of the mesh and at the internal (liquid-solid) interface. The external boundaries had temperature (time) dependant Dirichlet type conditions c_{L0} and c_0 on liquid and solid sides of the mesh, respectively. The Neumann type condition was implemented in the melting cell with a zero transport from the liquid to solid phase and vice versa.

The evolution of a melting cell is determined by both diffusion and melting processes. At the beginning of the time step the concentrations in the liquid and solid phases in the melting cell are equal to $c_{L\gamma}$ and $c_{\gamma L}$, respectively. At every time step we calculate first the changes in the concentration in the solid and liquid phases in the melting cell using the diffusion equation. The zero liquid-solid diffusion exchange condition imposed on the melting cell (cell number i, cells with numbers $< i$ are liquid) results in the following changes in solid, Δc_s, and liquid, Δc_l, concentrations produced by diffusion, adopted from Artemev et al. [11]:

$$\Delta c_s = \frac{\Delta t \cdot D_s \cdot (c_s[i+1] - c_s[i])}{\Delta x^2 \cdot (1 - f_l)}$$

$$\Delta c_l = \frac{\Delta t \cdot D_s \cdot (c_l[i-1] - c_l[i])}{\Delta x^2 \cdot f_l} \tag{6-3}$$

where $c_s[i]$ and $c_l[i]$ are concentrations in the solid and liquid phases in the cell number i. When these increments are added to the concentrations, the updated values $c_s^+[i]$ and $c_l^+[i]$ differ from the equilibrium conditions at the liquid–solid interface. $c_s^+[i]$ and $c_l^+[i]$ are used then to calculate the increase in fraction of liquid, Δf_l, using the condition that at the end of a time step equilibrium should be restored in the melting cell. The equation adopted from [11]:

$$\Delta f_l = \frac{(c_l^+[i] - c_{L\gamma}) \cdot f_l + (c_s^+[i] - c_{\gamma L})(1 - f_l)}{c_{L\gamma} - c_{\gamma L}} \tag{6-4}$$

At the end of a time step the fraction of liquid is updated using the calculated value of Δf_l, and equilibrium values are assigned to concentrations in liquid and solid. If Δf_l is such that an estimated fraction of liquid becomes larger than *1.0* then f_l is assigned the value *1.0* and melting is transferred into the next cell. The fraction of liquid in the next cell is calculated so that the equilibrium interface conditions are established there. At the very beginning of melting we need to initialize melting in the first cell in the system when the temperature becomes just higher than the eutectic temperature. We do this by using equations (6-1) and (6-2) to estimate the thickness of liquid produced in such a first melting step. After that the described combination of diffusion steps and melting steps is used. The simulation is effectively terminated when the temperature curve drops below the eutectic temperature after passing through the maximum.

6.2.1 Thermal, Microstructure and Stress Analysis

Thermal Analysis

The thermal analysis computes an *FEM* approximate solution to the transient *3D* energy equation (3-2), see chapter *III*, using *8*-node bricks with backward Euler time integration. It is solved on the current deformed geometry for each time step. Temperature dependent thermal conductivity and specific enthalpy, including the effect of latent heats of phase transformations were used,[11].

The heating affect of the arc is described by a Dirichlet boundary condition on nodes in the weld pool. The weld pool size, shape and position (as a function of time) are taken as data determined from experiment for each weld pass. The temperature in a weld pool is assumed to vary parabolically from the melting point at the boundary of the weld pool to a maximum temperature of the melting point plus *400 °K* at the weld pool centroid. Temperatures in the weld pool liquid/solid boundary were prescribed to *1526 °C (2780 °F, 1800 °K)*.The maximum temperature in the weld pool was prescribed to *1927 °C (3500 °F, 2200°K)*. The size and shape of the weld pool was estimated from macrographs contained in [8].

External boundaries of the structure which are assumed to be in still air, have a convection boundary condition with ambient temperature of *300 °K* and convection coefficient of *10 W/m² °K*. The pipe is filled with water and tilted slightly to avoid trapped air under the weld. On the internal surface of the water-filled pipe a coefficient in the range *500* to *5000 W/m² °K* and ambient temperature of *27 °C (80 °F, 300 °K)* was applied. For each weld for a fixed weld pool length, the thermal flux from the weld pool was computed for this range of convection coefficients on the internal pipe surface.

Assuming a weld efficiency factor of *0.65*, the computed net heat input from the arc was compared to the total heat input for the weld estimated from data taken from [8]. The convection coefficient that provided best agreement between weld nugget geometry and heat input data taken from [8] was *1500 W/m² °K*. Convections coefficients in the range *1000* to *2000 W/m² °K* did not substantially change the results.

Microstructure Analysis

The evolution of microstructure outside of the thin layer is computed using the methodology described in chapter *IV*. The microstructure evolution is assumed to be in equilibrium during heating, i.e., no super heating occurs. In austenite, grain growth begins after *Nb* and *V* carbo-nitrides dissolve and ceases when either delta ferrite forms on heating or ferrite, pearlite, bainite or martensite forms on cooling. Formation of ferrite, pearlite and bainite on cooling is modeled by *ODE*s, (see the general form in equation 4-21 and the special form for the austenite transformation to ferrite in equation 2-19). Martensite formation on cooling is modeled by the Koisten-Marburger equation.

Thermal Stress Analysis

The thermal stress analysis computes an *FEM* approximate solution to the conservation of momentum and mass. The loads, material properties and geometry are time dependent. Analogy to equation (5-12), chapter *V*, we say:

$$\nabla \cdot \sigma + f = 0$$

$$\frac{D\rho}{Dt} + \rho \nabla \cdot v = 0 \qquad (6-5)$$

where σ is the Cauchy stress, f is the body force and ρ is the density. Inertial forces are ignored.

For each time step, a displacement increment dU is computed using a Newton-Raphson iterative method in an updated Lagrangian formulation. The deformation gradient is in agreement with equation (5-6) for a time step $F = I + \nabla dU$. The Green-Lagrange strain increment for a time step is $d\varepsilon = (F^T F - 1)/2$. For details describing the finite strain theory, see chapter *V* and also Goldak et al [16].The deformation gradient is decomposed into elastic, initial and plastic deformation gradients, $F = F_{El} F_{Init} F_{Pl}$, where F_{Init} includes deformations due to thermal expansion and phase transformations.

For stress analysis, F_{Init} is made piecewise constant in an element. The increment in the 2nd Piola-Kirchoff stress is $d\sigma_{2PK} = Dd\varepsilon_{El}$ where D is the elasticity tensor. At temperatures below ~1100 °K rate independent plasticity is used. At temperatures above ~1700 °K a linear viscous constitutive model is used.

The residual for the Newton-Raphson iteration is:

$$\Re = P - \int B^T \sigma_{2PK} \qquad (6\text{-}6)$$

where B is the usual *FEM* matrix that maps nodal displacements to the strain at a point and P is the nodal vector of external loads. Each Newton-Raphson iteration computes a correction du to the current trial displacement dU by solving:

$$Kdu = -\Re \qquad (6\text{-}7)$$

where K is the global stiffness matrix. For each Newton-Raphson iteration, the linear equation (6-7) is solved to a tolerance of 10^{-9} in the energy norm. When the 1-norm of the residual $\Re$ is less than $10^{-4}P$, the time step is considered to have converged.

Coupling Thermal, Microstructure and Stress Analysis

The first time step computed the displacement, strain and stress due to applying the internal pressure, [10]. Then the weld pool was positioned in the deformed geometry for each subsequent time step. If a groove formed under the weld, the thermal analysis took into account this thinning of the pipe wall. When the arc was extinguished, the weld was allowed to cool to room temperature. The final time step returned the internal pressure to *1 atm*.

Test

Welding on a pressurized natural gas pipeline is simulated using computational weld mechanics to determine if *CWM* can provide useful estimates of the risk of burn through, [10]. The critical input data in addition to the internal pressure in the pipe, the geometry of the pipe, the size and shape of the weld pool including weld reinforcement, are the convection coefficient on the internal pipe surface and the temperature dependence of the viscosity of the pipe metal near the melting point.

The *CWM* analysis shows that above a critical state creep under the weld pool thins the pipe wall and forms a groove. When the pipe wall is thinned by the groove during welding, the internal pipe wall temperature increases under the weld pool during welding. This nonlinear interaction further increases the temperature on the internal pipe wall under the weld pool and further accelerates creep and the actual burn-through. The analyses shows significant thinning exactly in those welds that burned-through or were at high risk of burn-through. No significant thinning is predicted in those welds for which no significant thinning was reported experiments. We conclude that *CWM* can compute useful estimates of the risk of burn-through when welding on pressurized pipelines.

The thickness of the liquid layer and the concentration of carbon as a function of distance from the inner wall of the pipe have been computed. A *3D* transient temperature, displacement, stress and strain have been computed and coupled to the model for growth of the carburized layer.

The *3D* transient nonlinear thermal stress analysis used a viscoplastic material model and *8*-node bricks. Temperature dependent Young's modulus, Poisson's ratio, Yield strength, hardening modulus and viscosity are used. Rigid body modes in the pipe were constrained. The internal surface of the pipe was pressurized to *900 psig (6.2 MPa)*. The transient temperature field of the thermal analysis was applied to each time step. The displacement, strain and stress were computed for *40* time steps for each weld, i.e., the weld moved approximately *0.1 in (2.5 mm)* in each time step. The stress analysis was done on the full vessel.

Task Sequence

- Build an *FEM* mesh of the pressure vessel including the slots machined for welds and filler metal to simulate the welds described in the [8],

- For each repair weld, specify data of position, current, voltage, arc efficiency, welding speed, weld metal cross-section geometry and weld pool semi-axes lengths,
- For each weld compute the *3D* transient temperature, microstructure, displacement, strain and stress,
- Post-process results of each weld analysis to visualize the temperature, displacement, strain and stress. The deformation of the internal pipe wall under the weld was of particular interest,
- Compare the temperature and deformation computed with *CWM* with experimental temperature and deformation as estimated from macrographs in [8],
- Interpret the results of the analyses.

Structure and Materials

The pipe was *508 mm (20 in.)* diameter and *7.9 mm (0.312 in.)* wall *API 5LX-65 ERW*.

Table 6-1: Composition of HSLA steel

C	Ni	Si	V	Mo	W	Mn	Cr	Cu	P	Al
0.12	0.11	0.16	0.001	0.001	0.001	0.91	0.01	0.001	0.002	0.04

Table 6-2: Pipe material properties, from [8]

Designation	API-SLX65
Yield Strength	71.3 ksi (483 MPa)
Tensile Strength	89.8 ksi (621MPa)
Elongation %	29.4

Deposition of Welds

The welding consumables were Type *E7018* low hydrogen shielded metal arc electrodes. At the test weld location a flat grove was machined into the pipe and a single weld pass was deposited along the centerline of the machined area. The weld was made with an internal pressure of *6.2 MPa*.

All welds were made at an internal pressure of *900 psig (6.2 MPa)*. Tables 6-2 and 6-3 which are taken from [8] specify the data for each weld. The electrode size was *2.38 mm*, heat input was *0.79 KJ/mm* and travel speed was *2.1 mm/s*.

In [8] welds *A, B, D, G, H* and *I* were rated as ``safe'' welds; *C, J* and *L* were rated as marginal and weld *E* and *F* were rated as burn through. We only consider one weld from each class, i.e., welds *H, J* and *E*. For these three welds, the welding current was *80 amps*, the voltage *21 volts* and the electrode size *3/32 in.*

Table 6-3: Summery of Weld Data

Test ID	Wall Thickness (in.)	Heat Input (KJ/in.)	Travel Speed (in./min)
H	0.156	20	5.0
J	0.156	25	4.0
E	0.125	20	5.0

FEM Mesh

Figure 6-5 shows a composite mesh for a weld on a pressure vessel.

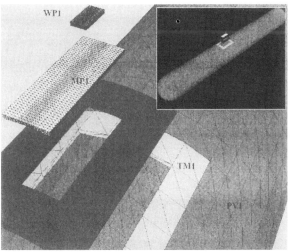

Figure 6-5: The various parts of the *FEM* mesh are shown. The smallest part, *WP1*, at the top of the figure, contains the weld pool.

The mesh for the pressure vessel, called *PV1*, had *3006* 8-node brick elements. For each weld, a fine mesh with *3456* elements, called *MP1*, was created for the region around the machined slot. Outside of *MP1*, a transition mesh with *768* elements, called *TM2*, and coarser transition mesh called *TM1* with *140* elements was made. Inside *MP1* an even finer mesh was created, called *WP1*, with *3840* elements.

Figure 6-6 shows the moving mesh near the weld pool region with the weld. Also filler metal is being added as the weld moves. The mesh *WP1* was also moved with the arc during the welding process.

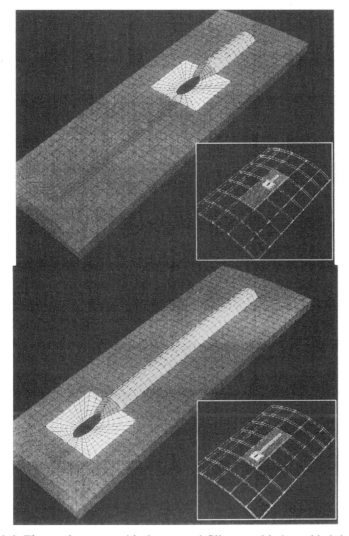

Figure 6-6: The mesh moves with the arc and filler metal being added during the welding process is shown.

Results and Discussion

Figure 6-7 shows the computed carbon concentration vs. distance from the internal wall at a point on the internal wall of the pipe.

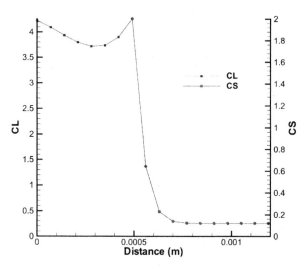

Figure 6-7: The carbon concentration distribution through the thickness of the liquid carburized film is shown at the point in time that the film growth stops. c_L is the carbon concentration in liquid, c_s is the carbon concentration in solid.

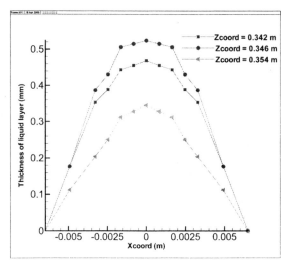

Figure 6-8: The thickness of the liquid layer as a function of distance from the weld centerline, *Xcoord.* =0, on the internal wall of the pipe. The thicker films are formed later in time and are downstream from the thinner films.

Figure 6-8 shows the thickness of the carburized film layer on several cross-sections vs. time. The dip in carbon concentration seen in Figure 6-8 corresponds to higher temperatures that shift the carbon concentration on the liquidus line to lower carbon values.

The following test results show that useful estimates of the risk of burn-through can be achieved by using the *FEM* methodologies.

Figure 6-9 shows some typical results of the thermal analysis.

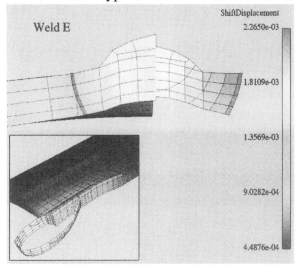

Figure 6-9: The *3D* transient temperature field at a point in time just before the arc was extinguished is shown. The isosurface is for *1225°K*. Note the groove in the pipe that formed under the weld is due to creep driven by the internal pressure in the pipe

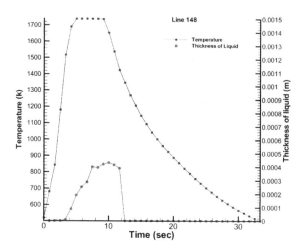

Figure 6-10: The temperature and thickness of the liquid layer versus time is shown for a point on the inside wall of the pipe.

Figure 6-10 shows a typical transient temperature distribution with geometry deformed by stress analysis.

The peak at the liquid-solid interface is caused by the temperature dropping to the eutectic temperature and that shifts the carbon level to the eutectic value. The results of our analysis are consistent with anecdotal evidence of a liquid carburized film forming on the inner wall of the pipe.

The thickness of the liquid layer and the concentration of carbon as a function of distance from the inner wall of the pipe have been computed. A 3D transient temperature, displacement, stress and strain have been computed and coupled to the model of growth of the carburized layer.

Figure 6-11 shows also results of the thermal analysis.

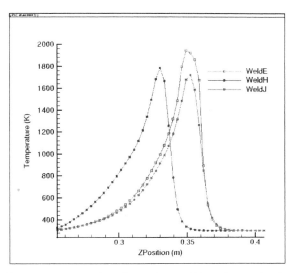

Figure 6-11: Temperature vs. *Z*, where time= *Z*/speed, on the internal surface directly below the weld path just before the arc was extinguished

The accuracy of the thermal analyses is considered to be limited primarily by the values chosen for the convection coefficient on the internal surface of the pipe. By setting the weld pool cross-section size and power input from data given in [8] and choosing an arc efficiency factor of *0.65*, we find the best agreement with data from [8] with the convection coefficient in the range *1000* to *2000* W/m^2 °*K*. We consider *1500* W/m^2 °*K* to be the best estimate of the value of the convection coefficient.

We first note that the stress state in the machined slot is rather different from that of a thin walled pipe with internal pressure. The thin walled slot actually deforms more like a bubble than a thin walled pipe. See Figure 6-12, curve A would be constant in a thin-walled pipe. The bending changes the stress distribution from that of a thin-walled pipe. Instead of a constant hoop stress, the hoop stress is higher on the outer surface where bending adds a tensile component and lower on the internal surface where bending adds a compressive component. Some of the hoop stress load is shunted around the slot as it would be shunted around a crack.

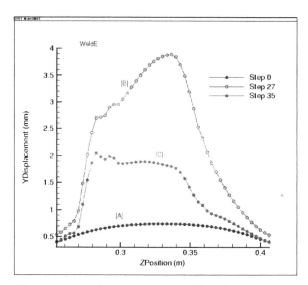

Figure 6-12: Displacement of internal surface directly below the weld path. A) After pressurizing the vessel but before starting weld (step *0*). B) Just after the arc was extinguished but before cooling to room temperature (step *27*). C) After the weld has cooled to room temperature (step *35*).

The agreement between experiment and *FEM* analysis for the total displacement of the deformed cross-sections in welds *H, J* and *E* is quite good. This suggests the *FEM* model is reasonably accurate. All welds that experiment showed had less risk of burn-through than weld *G* also had less thinning in the *FEM* analysis.

Kiefner [1 and 2] states that by the Battelle model, "It has been shown that for certain welding process, the level of *980 °C (1800 °F, 1255 °K)* at the inside surface is a safe upper limit for avoiding burn through". Figure 6-11 clearly shows that the difference in peak temperature on the internal wall of the pipe is not a sensitive measure of risk of burn through. This is not surprising because the peak temperature criterion ignores time and burn through is a time dependent phenomena. We argue that a much better model is obtained by *CWM* because, in addition to the peak temperature, it includes several additional parameters that are critical to the burn-through phenomena. These include the value of the internal pressure, the pipe diameter and wall thickness, the weld pool geometry and speed, convection coefficient, the area of the internal surface at temperatures above temperatures of *900°C,* and the time spent at high temperatures. The *CWM FEM* analyses of burn-through presented in this chapter do consider all of these factors. We argue that this is a significant advance in modeling burn-through when welding on pressurized pipelines, Figures 6-13 to 6-18.

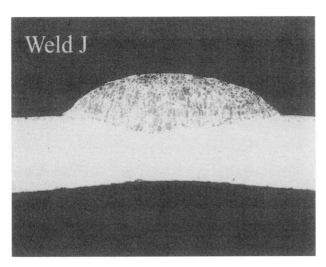

Figure 6-13: Experimental data for a marginal weld

It should be noted that almost no use has been made of adjustable or tuning parameters in the *FEM* analysis described in this section. Certainly there is some uncertainty in the convection coefficient on the internal surface of the pipe. However, the values used are in the range expected for a water filled pipe. There is also some uncertainty in the value of the high temperature viscosity for steel. Again the values used are in the range expected for thermal activated dislocation motion, i.e., high temperature creep in steels.

To simulate the actual burn-through, i.e., the formation of a hole, we conjecture that a dynamic analysis including inertial forces would be required.

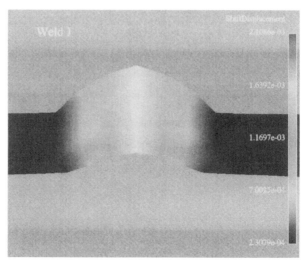

Figure 6-14: Computed deformed cross-section for a marginal weld. Cross-sections show reasonable agreement between the predicted and experimental deformation of a marginal weld.

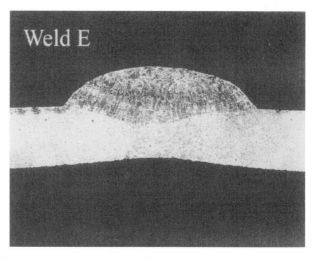

Figure 6-15: Experimental data for weld *E*, *A* burn-through weld

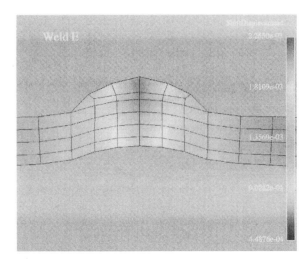

Figure 6-16: Computed cross-section for weld *E* cross-sections show excellent agreement between the predicted and experimental deformation of weld *E*.

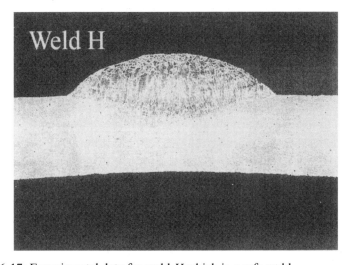

Figure 6-17: Experimental data for weld *H* which is a safe weld

If internal surface temperatures could be measured experimentally, the convection coefficient could be estimated more accurately. If deformation of the internal surface could be measured more accurately, then even higher accuracy comparisons could be made with *FEM* simulations.

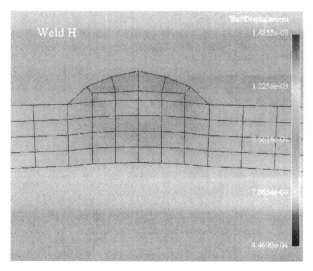

Figure 6-18: Computational data for weld *H* comparison with Figure 6-17 shows excellent agreement.

6.3 Hydrogen Diffusion in Welds

Hydrogen cracking is an important cause of failures in many welded structures. Given a sufficiently sensitive microstructure, a sufficient hydrogen concentration and a high enough tensile stress state, hydrogen cracking, sometimes called cold cracking, is likely to occur. Hydrogen cracking usually involves delays of minutes to years. If cellulosic electrodes are used, it is usual to assume that the weld pool is saturated with hydrogen. The solubility of hydrogen is highest in the weld pool and next highest in austenite and lowest in ferrite or martensite.

Therefore a *CWM* model that predicts hydrogen cracking must compute the diffusion of hydrogen in a transient temperature, microstructure and stress field. Hydrogen diffuses toward high hydrostatic stresses, i.e., tension. This is particularly important at low temperatures where stresses are higher and hydrogen diffusivity lower. In a weld, the diffusivity of hydrogen varies by several orders of magnitude with temperature and with phase changes. In addition, it requires a model for crack initiation and propagation. The simplest models simply assume the risk of cracking is too high if the hydrogen level exceeds a certain threshold level and the microstructure is sensitive to hydrogen cracking.

The hydrogen in a welded joint has three sources: from base metal and filling metals, from other consumables and from ambient air. The fraction of hydrogen from base metal and filling metal is essentially negligible because of their low hydrogen levels.

Microscopically, plastic deformations, dislocations, micro-voids, various inclusions, etc. may exist in a welded joint dispersively and vary from here to there. The activation coefficient will thus be greatly different from a "sound crystal". As a result, the diffusion does not follow Fick's law [24].

Recent work by Olson et al [30-32] clearly demonstrates the need for coupled weld metal and heat affected zone *(WM/HAZ)* approaches for dealing with the general problem of *HIC*, especially in regard to the high strength, low carbon materials. This is a consequence of the fact that the transformation temperature of the *WM* can be either higher or lower than the base material depending on relative filler and base metal compositions. It is said that the location of hydrogen cracks depends to a great extent on the relative martensite start temperatures of the two zones. If the martensite start temperature M_s of the *WM* is higher than of the *HAZ*, hydrogen accumulates in the coarse grained region. Austenite acts as something of a barrier to hydrogen movement into the *HAZ*, which accumulates in the boundary area causing the *HIC* problem at this location. On the other hand if the M_s of the *WM* is lower than that of the *HAZ*, there is less hydrogen accumulation in the sensitive boundary region and *HIC* is more likely in the *WM*. Bibby et al [28] propose an index based on the *WM/HAZ* difference in martensite start temperatures as follows:

$$\Delta M_s = M_{s_{WM}} - M_{s_{HAZ}} \tag{6-8}$$

Base Metal:

$$M_{s_{HAZ}} = 521 - 350C - 14.3Cr - 17.5Ni - 28.9Mn$$
$$- 37.6Si - 29.5Mo - 1.19Cr.Ni + 23.1(Cr + Mo).C \tag{6-9}$$

Weld Metal:

$$M_{s_{WM}} = 521 - 350C - 13.6Cr - 16.6Ni - 25.1Mn$$
$$- 30.1Si - 40.4Mo - 40Al - 1.07Cr.Ni + 21.9(Cr + 0.73Mo).C \tag{6-10}$$

For $\Delta M_s > 0$, the fracture is expected to occur in the *HAZ* and vice versa for $\Delta M_s < 0$.

6.3.1 Fundamental Equation

The driving force of hydrogen diffusion is the gradient of chemical potential, that is:

$$F = -\nabla\mu \tag{6-11}$$

where μ (cal/mol) is chemical potential, and:

$$\mu = \mu^\circ + RT \, ln\alpha \tag{6-12}$$

where μ° is the chemical potential when $\alpha=1$, namely standard chemical potential; α is the activity, R is the gas constant and T is the absolute temperature.

Generally the physical-chemical behavior of hydrogen during diffusion is represented by the activation coefficient γ and the product of the γ by the hydrogen concentration c, i.e., the so called activity α. That is $\alpha = \gamma c$.

Microstructural defects, such as grain-boundaries, dislocations and voids can be regarded as hydrogen "traps". When hydrogen diffuses and passes through them, it will be trapped partially. The "trapped" hydrogen does not participate in diffusion and only the "effective, diffusible" hydrogen, as the diffusing mass, must be considered in the diffusion equation. The activity reflects the very difference between the hydrogen concentration in the considered medium with and without these defects. The activity α may be called also "corrected hydrogen concentration" or "effective diffusible hydrogen concentration", [24].

Under a driving force F, the hydrogen moves with average velocity:

$$v = \frac{D}{RT}F \tag{6-13}$$

Then the hydrogen flux, *(mol/m²s)*:

$$J = vc = -\frac{Dc}{RT}\nabla\mu \qquad (6\text{-}14)$$

The hydrogen diffusivity has the general form $D = D_0 \exp[\frac{-Q}{RT}]$, m^2/s. The hydrogen diffusivity in austenite and ferrite as function of temperature [15]:

In austenite:

$$D = 1.1\times10^{-6}\exp(-\frac{41600}{8.31432\times T}) \qquad (6\text{-}15)$$

In ferrite:

$$D = 0.22\times10^{-6}\exp(-\frac{12100}{8.31432\times T}) \qquad (6\text{-}16)$$

By the law of conservation mass, if there is a hydrogen-source with an intensity of Q_H in certain infinitesimal element, the hydrogen concentration, c, should satisfy the following integral equation; n is the unit vector outwards and perpendicular to the surface:

$$\int_\Omega \frac{\partial c}{\partial t}d\Omega + \int_{\partial_2\Omega} J\cdot nds = \int_\Omega Q_H d\Omega \qquad (6\text{-}17)$$

From Gauss-Green integration law, the above Equation can be rewritten as:

$$\frac{\partial c}{\partial t} + \nabla\cdot J - Q_H = 0 \qquad (6\text{-}18)$$

The equation above is proposed by Zhang et al [3 and 24] as the general equation for hydrogen diffusion.

When hydrogen diffuses into a metal containing a stress field σ, there is an increase in the volume $\Delta V = V_h n$, where V_h is the volume increase due to one hydrogen atom or one mole of hydrogen atoms and n is the number of hydrogen atoms or number of moles of hydrogen added. The potential from the stress field that causes a hydrogen flux is, [18]:

$$\psi(\sigma) = V_h(\frac{\sigma_{ij}}{3} + \frac{\sigma\cdot\varepsilon}{2}) \qquad (6\text{-}19)$$

The flux associated with the stress field is:

$$j_\sigma = \frac{Dc}{RT}\nabla\psi \qquad (6\text{-}20)$$

The partial molar volume of hydrogen in ferrite, [20], is $V = 2.5\times10^{-6}$ *(m³/gram-atom)*, and in austenite, [21], is $V =1.75\times10^{-6}$ *(m³/gram-atom)*

The evolution of hydrogen content as a function of temperature in a stress field is:

$$\dot{c}-[\nabla\cdot D\nabla c+\nabla\cdot(\frac{Dc}{RT}\nabla\psi)]-Q_H = 0 \qquad (6\text{-}21)$$

If $c = 0$ or $\nabla\psi = 0$ on a boundary, then the hydrogen flux J on this boundary is zero. If $c \neq 0$ or $\nabla\psi \neq 0$ on a boundary, then the hydrogen flux on this boundary J is not zero. Sofronis et al [19] presents a solution for the hydrogen diffusion near a crack tip.

Streitberger and Koch have given a full account of the numerical solution of the diffusion equation with a stress term for a sharp crack tip under different loading conditions. The two-dimensional time-dependent drift-diffusion equation for a crack tip under mixed-mode loading and with tip as an ideal sink for solute atoms is solved by a finite difference method [25].

In all theoretical models it is assumed that the crack growth rate depends in a crucial way on the segregation rate of the embrittling solute to the crack tip region, whereas the segregation rate is governed by the concentration field $c = c(\vec{r},t)$ of the solute obeying the diffusion equation:

$$\frac{\partial c}{\partial t} = D\Delta c+\frac{D}{kT}(\nabla E)\nabla c+\frac{D}{kT}c\Delta E \qquad (6\text{-}22)$$

where D is the bulk or grain boundary diffusivity, T the absolute temperature, k the Boltzmann constant and $E = E(\vec{r})$ the relevant interaction energy between the external stress field and the impurity.

In an alternative model of dynamic embrittlement the equation (6-22) is considered for the somewhat simpler problem of the one-dimensional diffusion along a grain boundary ahead of a plastic crack tip. Only for the specific problem of transient hydrogen transport near a blunting crack has a full numerical analysis of the diffusion initial-boundary-value problem in conjunction with the elastic-plastic boundary value problem been carried out using finite element procedures [25].

With the assumption that the solute flow in the vicinity of the crack tip is dominated by the strong and long-ranged elastic interaction field and that the flow arising from random diffusion processes can be ignored at least for small time, the equation (6-22) for the harmonic field takes the explicit form following, [25 and 27]:

$$\frac{\partial c}{\partial t} = -\frac{AD}{2kTr^{3/2}}[\sin(\frac{\theta}{2} - \alpha)\frac{\partial c}{\partial r} - \cos(\frac{\theta}{2} - \alpha)\frac{1}{r}\frac{\partial c}{\partial \theta}]$$

$$A = (\frac{2}{9\pi})^{1/2}(1+v)K\Delta V \qquad -\pi < \theta < \pi, r << a \qquad (6\text{-}23)$$

where $K = \sigma_1(\pi a)^{1/2}$ is the stress intensity factor.

The complete drift-diffusion equation (6-22), which governs the migration of point defects around the crack tip under the action of the harmonic interaction potential (6-23), takes the form:

$$\frac{\partial \bar{c}}{\partial \bar{t}} = \frac{\partial^2 \bar{c}}{\partial \bar{r}^2} + \frac{1}{\bar{r}}\frac{\partial \bar{c}}{\partial \bar{r}} + \frac{1}{\bar{r}^2}\frac{\partial^2 \bar{c}}{\partial \theta^2}$$

$$-\frac{Q\sin(\alpha)}{\bar{r}^{3/2}}(\sin(\frac{\theta}{2} - \alpha)\frac{\partial \bar{c}}{\partial \bar{r}} - \cos(\frac{\theta}{2} - \alpha)\frac{1}{\bar{r}}\frac{\partial \bar{c}}{\partial \theta}) \qquad (6\text{-}24)$$

where $\bar{c} = \frac{c}{c_0}, \bar{r} = \frac{r}{R}, \bar{t} = \frac{Dt}{R^2}$ are the scaled variables and the parameter Q is defined by:

$$Q = \frac{A}{2kTR^{1/2}} \qquad (6\text{-}25)$$

R is an additional length scale that defines a large cylindrical region around the crack tip where the numerical solution of equation (6-24) is performed. The initial and boundary conditions are respectively:

$$\bar{c}(\bar{t}, \bar{r} = 1, \theta) = 1$$
$$\bar{c}(\bar{t}, \bar{r} = \bar{a}, \theta) = 0 \qquad (6\text{-}26)$$

Furthermore, at the crack faces $(\theta = \pm\pi, \bar{r} > \bar{a})$ the tangential component of the solute flux density is required to be zero:

$$\bar{j}_\theta = \frac{\partial \bar{c}}{\partial \theta} + \frac{Q\bar{c}}{\bar{r}^{1/2}}\sin \alpha \cos(\frac{\theta}{2} - \alpha) = 0 \text{ for } \theta = \pm\pi \qquad (6\text{-}27)$$

The zero-flux boundary condition (6-27) is required to model impermeable crack faces.

6.3.2 Preheat prediction

Several investigators have developed relationships for predicting preheat levels to avoid the *HAZ* hydrogen cracking problem. Bibby et al [28] present a review of the predictive methods to assist in managing hydrogen in welding applications.

The following method may be used to determine the necessary preheat temperature for steel welding. This method is based on Stout H Slit or Tekken Test experimental results, [29]:

1- Obtain the carbon equivalent of the steel to be welded using equation (6-28c)

2- Obtain the estimate of the hydrogen content of the process.

3- Determine K_t using the chart in Figure 6-19, and the weld stress σ_w using with the help of the restraint intensity in Figure 6-20.

4- Calculate *CI* using the values of $CE, H_{JIS}, K_t, \sigma_w$ and equation (6-28b)

5- Calculate $(t_{100})_{cr}$ from *CI* using equation (6-28a) or obtain from Figure 6-21.

6- Finally select the preheating temperature, taking into account h, $2b$ and Q/v, so that the $t_{100} > (t_{100})_{cr}$, Figures 6-22 and 6-23.

The chart method for determining *HAZ* preheat to avoid *HIC* described above, adopted from Bibby et al [28], is convenient but less general than a computer managed system.

The Slit Test is a self-restrained cracking test that has been applied to pipeline girth welds. In the Slit Test, welds are performed at a sequence of preheat temperatures. The lowest preheat temperature at which hydrogen cracking does not occur is the critical preheat temperature. To use this method both the critical and actual weld cooling times must be calculated.

$$t_{100} \geq (t_{100})_{cr} \tag{6-28}$$

$$(t_{100})_{cr} = \exp(67.6CI^3 - 182.0CI^2 + 163.8CI - 41.0) \tag{6-28a}$$

$$CI = CE_N + 0.15 \log H_{JIS} + 0.30 \log(0.017 K_t \sigma_w) \tag{6-28b}$$

$$CE_N = C + A(C).\{\frac{Si}{24} + \frac{Mn}{6} + \frac{Cu}{15} + \frac{Ni}{20} + \frac{Cr + Mo + Nb + V}{5} + 5B\} \tag{6-28c}$$

$$A(C) = 0.75 + 0.25 \tanh\{20(C - 0.12)\} \tag{6-28d}$$

The carbon equivalent is easily determined from the chemical composition of the base material. The weld pool diffusible hydrogen level H_{HIJ} can be estimated as follows [28 and 34]:

1.0 ml/100g for low hydrogen *GMA* situations;

3.5 ml/100g for *SAW* and low hydrogen *SMAW*;

15.0 ml/100g for rutile/acid electrodes and;

30-40 ml/100g for cellulosic field electrodes;

The joint configuration stress concentration factor varies from a low of about *1.5* for a full *V* root pass to about *7* for a single bevel mid-thickness root pass, Figure 6-19.

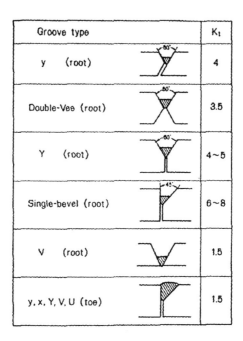

Groove type		K_1
y (root)		4
Double-Vee (root)		3.5
Y (root)		4~5
Single-bevel (root)		6~8
V (root)		1.5
y, x, Y, V, U (toe)		1.5

Figure 6-19: Stress concentration factors at root and toe weld positions, adopted from Bibby et al [28].

The local stress at the toe of a root weld, σ_w, is $K_t s_w$ where s_w is the stress across a weld *(MPa)* . The mean stress across a weld is given by $s_w = 0.05 R_f$ where $R_f < 20 s_y$; s_y is the yield stress *(MPa);* $s_w = s_y + 0.0025(R_f - s_y)$, where $R_f > 20 s_y$. Restraint R_f *(N/mm mm)* is by definition the force per unit length necessary to expand or contract the joint gap by a unit length. It is evaluated by the expression $R_f = 71 r_f [\arctan(0.017h) - (h/400)^2]$. The restraint coefficient r_f *(MPa/mm)* is a fundamental factor that assumes a two dimensional stress rate. The restraint R_f is thickness dependent, which accounts for three-dimensional effects. r_f can be estimated by knowing that it varies from a low of about *400 MPa/mm* for normal restraint situations to *700 MPa/mm* for high restraint situations. By combining the carbon equivalent, the diffusible hydrogen level and the local stress level, a cracking index *CI* can be calculated and from that the critical time t_{cr}.

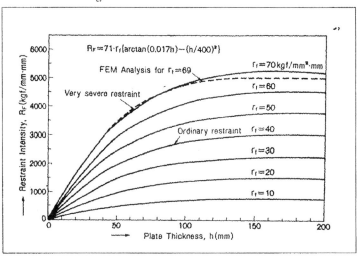

Figure 6-20: Relationship of restraint intensity to plate thickness, adopted from Bibby et al [28]

It is then necessary to calculate the actual weld cooling time. The cooling time can calculated from the following time-temperature relationship, adopted from Yurioka et al [35] and Bibby et al [28]:

$$T = T_\infty + \{\frac{(Q/v)\lambda}{\kappa h(cm)} \frac{1}{\sqrt{4\pi\lambda t}} + (T_o - T_\infty)\} \cdot \exp(-\frac{aS\lambda}{\kappa V} \cdot t) \qquad (6\text{-}29)$$

where Q/v *(J/mm)* is the heat input per unit length of weld, the thermal conductivity κ =0.06 *J/mm°C* is recommended, the thermal diffusivity λ=14.6 *mm²* */°C,* convection coefficient $a = 1{,}5 \times 10^{-5} J / mm^2 s^2 {}^\circ C)$, S is the surface area of the plate *(mm²)*, *V* is the volume of the plate *(mm³)*, *h* is plate thickness *(mm)*, *t* is time *(s)*, *b* is the half width of the electrical strip preheater *(mm)*, T_∞ is the ambient temperature *(°C)*, T_o is the preheat temperature *(°C)*.

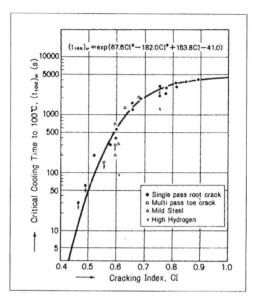

Figure 6-21: Relationship between critical cooling time to *100°C* and cracking index, adopted from Bibby et al [28]

If the cooling time is less than the critical, a higher preheat is used and the calculation is repeated until the inequality of equation (6-28) is satisfied. Where preheat is local (e.g., strip electrical heaters of width 2b) the preheat temperature is calculated according to the following:

$$T_o = T_\infty + \frac{q.b}{4\kappa} F(\beta_o)$$ (6-30)

$$F(\beta) = \frac{1}{2\beta^2} erf(\beta) + \frac{1}{\beta\sqrt{\pi}} \exp(-\beta - \{1 - erf(\beta)\}$$ (6-30a)

$$\beta_o \equiv b / \sqrt{4\lambda t}$$ (6-30b)

where *q'* is the strip preheater strength *(J/mm s)*, *q=q'/h* is the preheater strength per unit of cross sectional area, typically *0.05-0.2 (J/mm² s)* and t_{ph} is the preheating time *(s)*.

Substituting the local preheat temperature into the following relationship provides a time-temperature relationship from which a cooling time can be calculated

$$T=T_\infty+\{\frac{(Q/v)\lambda}{\kappa h(cm)}\frac{1}{\sqrt{4\pi\lambda t}}+(T_o-T_\infty)\frac{F(\beta_1)-F(\beta_2)}{F(\beta_o)}\}\cdot\exp(-\frac{2a\lambda}{\kappa h}\cdot t) \quad (6\text{-}31)$$

$$\beta_1 \equiv b/\sqrt{4\lambda(t+t_{ph})} \quad (6\text{-}31a)$$

$$\beta_2 \equiv b/\sqrt{4\lambda t} \quad (6\text{-}31b)$$

The preheat is adjusted in equation (6-30) until the cooling time calculated in equation (6-31) satisfies the inequality of equation (6-28).

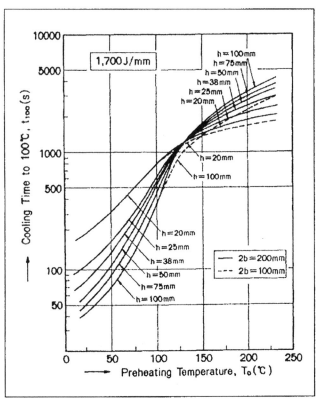

Figure 6-22: Relationship between cooling time to *100°C* and preheating temperature *(Q/v=1700 J/mm)*, adopted from Bibby et al [28].

(Q/v=1700J/mm).

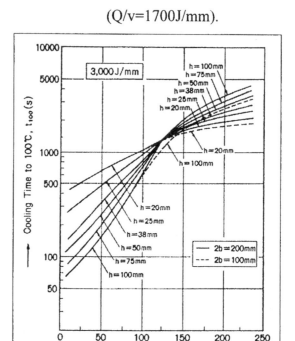

Figure 6-23: Relationship between cooling time to *100°C* and preheating temperature *(Q/v=3000 J/mm)*, from [28]

6.3.3 Computational Analysis

By *3D-FEM* analysis for hydrogen diffusion, the heat transfer, stress-strain analysis and transient microstructure distribution must be coupled. This *FEM* formulation leads to asymmetric equations. Other than that the formulation is straight forward. The value of temperature at the nodes is obtained by solving the associated energy problem. The microstructure is evaluated at the nodes by solving the associated ordinary differential and algebraic equations for the evolution of austenite grain size and the decomposition of austenite into ferrite, pearlite, bainite and martensite. The value of ψ at the nodes is obtained by solving the associated stress problem. Given these values, the value of hydrogen at the nodes is obtained by solving the solute diffusion problem.

For each time step, the variations of the temperature, stress, strain and the volumetric fractions of different phases should be obtained, so that, the physical-chemical "constants" in every phase, such as activation coefficient, diffusion coefficient, transfer coefficient, solubility and so on, of every element can be calculated. The concentration of the "effective, diffusible" hydrogen, c^*, may be obtained through the following mass diffusion equation, adopted from [3 and 24]:

$$[K]\{c^*\}+[R]\frac{\partial}{\partial t}\{c^*\}=\{F\} \tag{6-32}$$

where matrix $[K]$, $[R]$ are related to the activation γ and diffusion D coefficients; $\{c^*\}$ is the concentration matrix; $\{F\}$ is the hydrogen flow matrix. The obtained c^* multiplied by β makes c, the actual hydrogen concentration, where $\beta=1/\gamma$. According Dubios [26] $\beta=1+\varepsilon_{pl}$ where ε_{pl} is the plastic strain produced in welding.

Goldak et al [15] refers to the computational analysis of the Slit test and compares the results to experimental data. The *3D* transient temperature, microstructure, stress and strain fields in the Slit Test are computed. In addition, the *2D* transient hydrogen concentration field is computed on the central cross-section. The hydrogen diffusion model assumes the weld pool has a prescribed hydrogen level. In addition the hydrogen flux due to gradients in the stress field is included. The *3D* transient temperature, microstructure and stress are coupled to the *2D* hydrogen field.

A mesh for a hydrogen diffusion simulation is shown in Figure 6-24.

Figure 6-24: Mesh for hydrogen diffusion simulation

Figure 6-25 and Figure 6-26 shows the important effect of hydrostatic stress and on the hydrogen distribution.

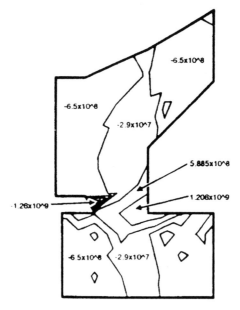

Figure 6-25: Hydrostatic stress distribution at temperature *100°C*

The compressive stress at the bottom of the weld tends to repel the hydrogen. The formation of two peaks in the hydrogen distribution near the center of the weld also appears to be associated with the hydrostatic stress distribution, Figure 6-25.

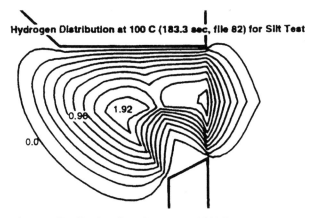

Figure 6-26: Hydrogen distribution for Slit Test at *100°C*

Figure 6-27 shows that the effective plastic strain rate approaches its maximum value quite soon after welding and before hydrogen peaks. This could be useful as it suggests the critical condition is reached when hydrogen peaks (since strain is then constant). It also suggests the time at which hydrogen cracking could occur.

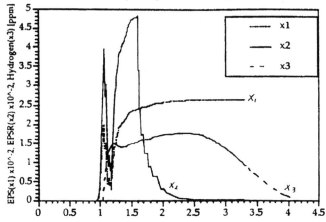

Figure 6-27: At the point with maximum hydrogen levels at *50°C* in slit test [15], the effective plastic strain *(x1)*, effective plastic strain rate *(x2)* and evolution of hydrogen *(x3) ppm* are plotted vs. log (time *s*)

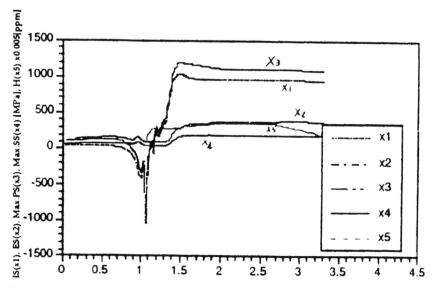

Figure 6-28: At the point with maximum hydrogen levels at *50°C* in Slit Test, the hydrostatic stress x1, effective stress x2, principal stress x3, maximum shear stress x4 and evolution of hydrogen x5×0.005 *ppm* are plotted vs. log time *(s)*.

The numerical study by Streitenberger and Koch sheds some light on stress-driven diffusion processes of solute near stress concentrators and should be useful for the interpretation of dynamic embrittlement and impurity induced fracture processes at elevated temperatures.

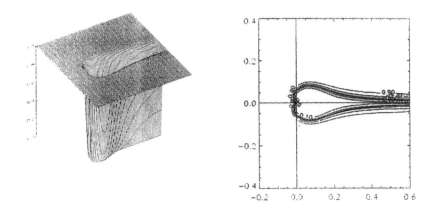

Figure 6-29: Concentration distribution (left) and lines of constant concentration (right) in the vicinity of a crack tip in the opening mode for a drift parameter $Q=0.5$ and after the reduced time $\bar{t} = 0.01$ in the limit of pure drift, adopted from Streitenberger and Koch [25].

In Figures 6-29 and 6-30 the calculational power and accuracy by Streitenberger et al [25] numerical scheme is tested for the limiting cases of pure drift and pure random diffusion, respectively, in the vicinity of a mode-I loaded crack tip.

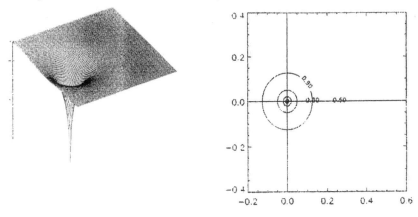

Figure 6-30: As for Figure 6-29 but in the limit of pure random diffusion $Q=0$, adopted from Streitenberger and Koch [25]

In the Figures 6-29 and 6-30 a three dimensional plot of the solute concentration field and the lines of constant concentration after 10000 time steps are displayed. As illustrated, the typical features of the pure-drift approximation, namely the discontinuity in the solute concentration separating the depleted from the non-depleted region on both sides of the crack faces and the shape of the characteristics, are well reproduced by the Streitberger/Koch [25] numerical solution.

References

1. Kiefner J. Effects of flowing products on line weld ability, Oil and Gas Journal, pp 49-54, July 18 1988
2. Kiefner J. and Fischer R. Models aid in pipeline repair welding procedure, Oil and Gas Journal, pp 41-547, March 7 1988
3. Zhang C. and Goldak J.A. Computer simulation of 3D hydrogen diffusion in low alloy steel weldment, IIW Doc. IX-1662-92
4. Liubov B. Em Diffuzionnye protsessy v neodnorodnykh tverdykh sredakh, (diffusion processes in inhomogeneous solids), Moscow Nauka, 1981.
5. Kurz W. and Fisher R. Fundamentals of Solidification, Trans. Tech. Publications, Switzerland-Germany-UK-USA, 1989
6. Umantsev A.R., Vinogradov V.V. and V.T. Borisov V.T. Mathematical model of growth of dendrites in a super cooled melt. Sov. Phys. Crystallogr. 30: # 3, pp 262-265, 1985
7. Artemev A. and Goldak J.A. Cellular simulation of the dendrite growth in Al-Si alloys, Canadian Metal. Quart, Vol.36, pp 57-64, 1997
8. PRC-185-9515, PRC Project: Repair of Pipelines by Direct Deposition of Weld Metal - Further Studies, August 3 1995
9. Henwood C. Bibby M.J., Goldak J.A. and Watt D.F. Coupled transient heat transfer microstructure weld computations, Acta Metal, Vol. 36, No. 11, pp 3037-3046, 1988
10. Goldak J.A., Mocanita M., Aldea V., Zhou J., Downey D., Dorling D. Predicting burn-through when welding on pressurized pipelines, Proceedings of PVP'2000, 2000 ASME Pressure Vessels and Piping Conference, Seattle Washington USA, July 23-27 2000
11. Artemev A., Goldak J. and Mocanita M. Carburization during welding on pressurized natural gas pipelines, ICES'2K, Los Angles CA USA, Aug. 21-25 2000
12. Goldak J., Breiguine V., Dai N., Zhou J. Thermal stress analysis in welds for hot cracking, AMSE, Pressure Vessels and Piping Division PVP, Proceeding of the 1996 ASME PVP Conf., July 21-26, Montreal.

13. Goldak J.A., Breiguine V., Dai N., Zhou J. Thermal stress analysis in welds for hot cracking, Editor H. Cerjak H., 3rd Seminar, Numerical Analysis of Weld ability, Graz, Austria, Sept. 25-26 1995.
14. Watt D. F., Coon L., Bibby M. J., Goldak J.A. and Henwood C. Modeling microstructural development in weld heat affected zones, Acta Metal, Vol. 36, no 11, pp 3029-3035, 1988.
15. Goldak J. A., Gu M., Zhang W., Dai N., Artemev A., Gravellie B., Glover A. and Smallman C. Modeling the slit test for assessing sensitivity to hydrogen cracking, International Conference Proceedings on modeling and Control of Joining Processes, Orlando Florida USA, Dec. 8-10 1993
16. Goldak J.A., Breiguine V., Dai N., Hughes E. and Zhou J. Thermal Stress Analysis in Solids Near the Liquid Region in Welds. Mathematical Modeling of Weld Phenomena, 3 Ed. By Cerjak H., The Institute of Materials, pp 543-570, 1997
17. Crank J. Mathematics of Diffusion, 1975
18. Larche F.C. and Cahn J.W. The effect of self-stress on diffusion in solids, Act Met., Vol. 30, pp 1835-1845, 1982
19. Sofronis P. and Birnbaum H.K. Hydrogen enhanced localized plasticity: A mechanism for hydrogen related fracture, Fatigue and fracture of aerospace structural materials, ASME, Vol. 36, pp 15-30, 1993
20. Bai Q., Chu W. and Hsiao C. Partial molar strain field of hydrogen in alpha-Fe, Scripta Metallurgica, Vol. 21, pp 613-618, 1987
21. Alefeld G. and Volkl J. Hydrogen in Metals I; Basic Properties, Springer Verlag, Berlin Germany, 1978
22. Yurioka N., Suzuki H. and Ohshita S. Determination of necessary preheating temperature in steel welding, welding Journal AWS, Vol. 62, No. 6, pp 147s-154s, 1983
23. Bibby M.J., Goldak J.A., Jefferson I. and Bowker J. A methodology for computing heat affected zone hardness, microstructure and preheat temperature, Computer Technology in Welding, Cambridge UK, June 8-9 1988
24. Zhang C. Numerical simulation of the hydrogen accumulation at the microscopic scale in a low-alloy steel weldment, 3rd International Seminar on Numerical Analysis of Weldability, Graz Seggau Austria, Sept. 24-27 1995
25. Streitenberger P. and Koch M. Stress-driven diffusion of impurities near crack-like singularities and mechanisms of dynamic intergranular embrittlement, 5th International Seminar on Numerical Analysis of Weldability, Graz Seggau Austria, Oct. 1999
26. Dubois D. et al: Numerical simulation of a welding operation: Calculation of residual stresses and hydrogen diffusion, Int. Conf. on Pressure Vessel Tech. San Francisco, Sept. 9-14 1984.
27. Rauh H., Hippsley A. and Bullogh R. The effect of mixed-mode loading on stress-driven solute segregation during high-temperature brittle intergranular fracture, Acta Met., 37(1), pp 269-279, 1989

28. Bibby M.J., Yurioka N., Gianetto J.A. and Chan B. Predictive methods for managing hydrogen in welding applications; Hydrogen Management for Welding Applications Proceedings of International Workshop, Ottawa Canada, October 6-8, 1998

29. Yurioka N., Suzuki H., Ohshita S. and Saito S. Determination of necessary preheating temperature in steel welding, Welding Journal, Vol. 62, No. 6, pp 147s-153s, 1983

30. Wang W.W., Lui S. and Olson D.L. Consequences of weld under matching and over matching; Non-uniform hydrogen distribution, Materials Engineering Proceedings of the 15th International Conference on Offshore Mechanics and Artic Engineering, ASME Part 3 (of 6), Vol. 3, pp 403-409, 1996

31. Wang W.W., Wong R., Liu S. and Olson D.L. Use of martensite start temperature for hydrogen control, Welding and Weld Automation in Shipbuilding Proceedings, TMS Materials Week 95, Cleveland, pp 17-31, 1995

32. Olson D.L., Liu S., Wang W., Pieters R.R.G.M. and Ibarra S. Martensite start temperature as a weldability index, Proceedings of 4th International Conference on Trends in Welding Research, ASM International, Materials Park, Ohio USA, pp 615-620, 1996

33. Olson D. L., Maroef I, Lensing C., Smith R.D., Wang W.W., Lui S., Wilderman T. and Eberhart M. Hydrogen management in high strength steel weldments, Proceedings of Hydrogen Management in Steels Weldments, Melbourne Australia, Pub. Defence and Technology Organization and Welding Technology Institute of Australia, ISBN 0 7311 0809 4, pp 1-20, 1997

34. Yurioka N and Suzuki H. Hydrogen assisted cracking in C-Mn and low alloy steel weldments, International Materials Reviews, Institute of Metals and ASM International, Vol. 35, No.4, pp 217-250, 1990

35. Yurioka N., Okumura M., Ohshita S. and Shoja S. On the method of determining preheating temperature necessary to avoid cold cracking in steel welding, IIW Doc. XI-E-10-81, 24 pgs, 1981

36. Armero F. and Love E. An ALE finite element method for finite strain plasticity, ECCM European Conference on Computational Mechanics, Cracow Poland, June 26-29 2001

Chapter VII

Welded Structures and Applications of Welding in Industrial Fields

7.1 Introduction and Synopsis

A welded structure is usually designed by a structural engineer who designs the structure but only specifies the requirements of the welded joints. A welding engineer, not the structural engineer, specifies the weld procedure for each joint to meet the requirements specified by the structural engineer.

Next a production engineer specifies the sequence of welds, i.e., the order in which welds are made and where each pass in each weld joint starts and stops. Various inspection methods such as ultrasonic, x-ray and magnetic particle can be used to look for defects in the welds. Finally the structure may be stress relieved and/or pressure tested.

This is a very brief summery of the design and fabrication process for a welded structure.

To date this process relies primarily on experience and testing. Computational Mechanics has been used primarily to analyze short single pass welds in test coupons. It has rarely been applied to analyze welds in real welded structures. There are several reasons for this:

1. The Numerical methods for Computational Welding Mechanics *(CWM)* have only reached sufficient maturity to contemplate the analysis of welded structures in the last decade.

2. The numerical methods for *CWM* have tended to be too computational intensive to analyze more than a single pass weld longer than one meter.

3. The preparation of input data to do the analysis, particularly creating the mesh, has been so complex and hence difficult that it has discouraged attempts to analyze complex welded structures.

Analyses of single weld joints in real structures are beginning to appear [1]; however, the major obstacle for using the simulations in industrial practice is the need for material parameters and the lack of expertise in modeling and simulation. The Finite Element Method is the most important tool used in simulating the thermomechanical behavior of a structure during welding. It is a general tool but may be computer demanding, [32]

By giving designers the capability to predict distortion and residual stress in welds and welded structures, they will be able to create safer, more reliable and lower cost structures. This capability is expected to become available to industry in the near future.

Real-time *CWM* will become feasible. The reason is that the speed of computers increases by a factor of roughly *1.7* times each year. In the past *18* years the speed of computers has increased roughly *2,000* times. In the next *18* years the speed of computers is expected to increase at least as rapidly. Clearly even if one does nothing to improve software for *CWM*, someday real time *CWM* would become feasible simply due to the increasing power of available computers.

This chapter presents a methodology for the computational welding mechanics analysis of welded structures with complex geometry and many weld joints and such that each weld joint could have more than one weld pass. The methodology is intended to be used to analyze the fabrication of complex welded structures such as ships, automotive, heavy equipment and piping networks. This methodology promises to facilitate research into the effect of welding sequence on a structure, the effects of interaction between welds on a structure and the effect of welds on the behavior of a structure, particularly the life of a structure.

The methodology supports distributed design in that the structure to be welded, the weld joints, the weld procedure and the sequence of welding the weld joints can be specified independently by a structural, a welding and a production engineer. It also supports integration and team centered design.

7.2 Weld Procedure

A weld procedure is a specification for a weld joint defined on a cross-section of the weld joint. It specifies the number of weld passes, the position and shape of the nugget for each pass and the details of the weld process for each pass. The details include the type of weld process, type and size of consumables, welding current, voltage, welding speed, preheat, interpass temperature, the qualifications of the welder, etc.

A weld procedure that has been tested and proved to meet the standards of a testing organization is called a qualified weld procedure. Companies can have libraries with tens or hundreds of qualified weld procedures.

The weld procedure is usually not designed for a particular weld joint in a particular structure. A weld procedure is usually designed to be used for a fairly wide class of weld joints. In terms of object oriented software engineering, weld procedures should be implemented as an abstract class and each instance of a weld procedure should be instance of that class. Each weld joint must specify one weld procedure but no weld procedure should specify a weld joint. Table 7-1 shows an example of a weld procedure.

Table 7-1: An example of a weld procedure, from [2]

SHIELDED METAL-ARC (MANUAL)

Position: Horizontal Weld Quality Level: Code Steel Weldability: Poor	⊢ 1/2" 3/8"	⊢ 1/2 – 5/8" 3/8 – 1/2"		⊢ 3/4" 5/8"	⊢ 1" 3/4"
Weld Size, L (in.)	3/8*	3/8	1/2	5/8	3/4
Plate Thickness (in.)	1/2	1/2	5/8	3/4	1
Pass	1	1 – 2	1 – 2	1 – 3	1 – 4
Electrode Class	E7028	E7028	E7028	E7028	E7028
Size	1/4	7/32	1/4	1/4	1/4
Current (amp) AC	390	335	390	390	390
Arc Speed (in./min)	7.5 – 8.5	11.5 – 12.5	9.0 – 10.0	9.0 – 10.0	8.0 – 9.0
Electrode Req'd (lb/ft)	0.483	0.483	0.819	1.28	1.82
Total Time (hr/ft of weld)	0.0250	0.0333	0.0422	0.633	0.940

Preheat may be necessary depending on plate material.

If a capability of predicting the weld pool geometry is available, it could be used to predict the weld pool geometry for each weld pass.

A linear correlation between the depth and the length of the weld pool is found in laser beam welding experiments with varied laser beam power. Approximately *50-90%* of the weld pool length (increasing with welding speed) results from conductive heat transport (with the fusion zone convexity contributing approximately *20-30%*). The remaining *50-10%* of the weld pool length (decreasing with welding speed) results from convective heat transport, [26].

The prescribed temperature distribution function proposed by Goldak, (see chapter *III*) has been developed to model weld pools with more complex geometry.

Fusion welding processes, in which the metal parts are heated until they melt together, can be performed with or without the addition of filler material. Arc welding, electron beam welding and laser welding belong to this category of welding processes. Radaj [31] gives more insight into the different processes and phenomena of which one should be aware.

7.3 Weld Joint

A weld joint usually joins two or more parts. In any case we assume that the position of any weld joint in space can be associated with a curve in *3D* space. This curve is parameterized by distance and hence has a start and stop point. At each point on this curve, a tangent vector *r* exists and a normal vector *s* is specified. In addition a third basis vector *t* is specified that is orthogonal to *r* and linearly independent of *s* but need not be orthogonal to *s*. Hence at each point on the curve the three basis vectors *(r, s, t)* define a curvilinear coordinate system. The basis vectors *(s, t)* define a *2D* curvilinear coordinate system at each point on the curve in a plane normal to *r*, i.e., the cross-section to the curve.

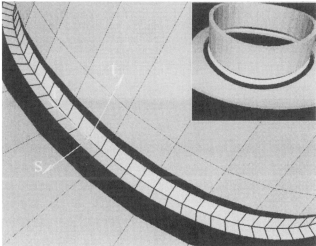

Figure 7-1: An example of a curvilinear coordinate system for a weld Tee-joint.

The weld procedure can be imagined to slide along this curve so that at any point on the curve, the weld procedure lies in the plane normal to *r*. Figure 7-1 shows an example of such a curvilinear coordinate system for a weld Tee-joint and Figure 7-2 shows an example of the geometry of a Tee-joint with three weld passes.

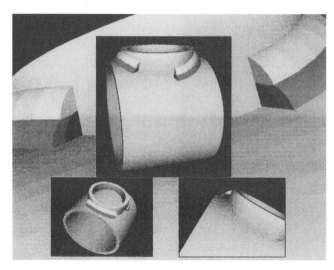

Figure 7-2: The two parts of a Tee-joint to be welded and a possible weld procedure with three weld passes.

7.4 Analysis of a Weld Structure

Figure 6-5, chapter *VI*, shows a composite mesh for a weld on a pressure vessel. These meshes do not always conform; i.e., nodes and element faces on either side of the interface need not match.

Continuity of temperature or displacement fields across the interface can be maintained by constraints. These constraints are imposed automatically by the code. Composite meshes make it much easier to mesh complex structures because each part of the structure can be meshed independently and the mesh is not required to maintain continuity of the mesh. In some cases, this can greatly reduce the work needed to mesh a structure for *CWM*. However, evaluating the cost of constraints increases the cost of analyses.

Figure 6-6, chapter *VI*, shows the moving mesh near the weld pool region with the weld. Also filler metal is being added as the weld moves.

Figure 7-3 shows a Tee-joint mesh including filler metal for the three passes.

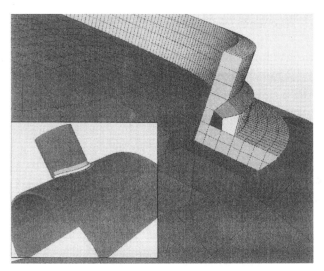

Figure 7-3: The weld Tee-joint mesh including filler metal for the three passes

Figure 7-4 shows the temperature near the weld pool with three temperature isosurfaces.

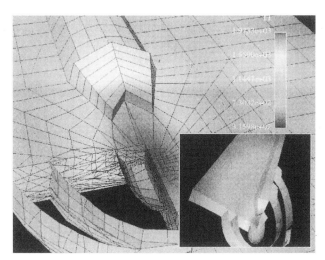

Figure 7-4:.The temperature near the weld pool with three temperature isosurfaces

Somewhat similar efficiencies could be achieved by adaptive meshing, which automatically refines and coarsens a finite element mesh [13].

Adaptive meshing has been used in some studies, for Lagrangian meshes, so as to utilize the degrees of freedom in the computational model better by concentrating them to regions where large gradients occur. Siva Prasad and Sankaranarayanan [19] used triangular constant strain elements. McDill et al. [20, 21, 22 and 23] implemented a graded element that alleviates the refining/coarsening of a finite element mesh consisting of quad elements, in two dimensions, or brick elements, in three dimensions. It has been used successfully for three-dimensional simulations [17] where the computer time was reduced by *60%* and with retained accuracy, Figure 7-5, for the electron beam welding of a copper canister.

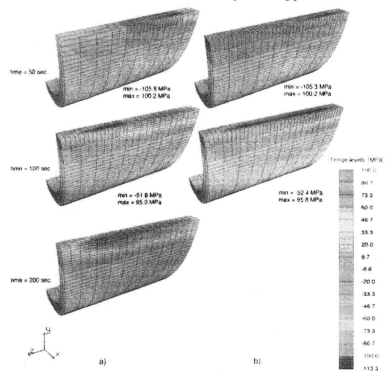

Figure 7-5: Axial stresses (a) with remeshing at *10, 50, 100* and *200 sec.* and (b) without remeshing at *50* and *100 sec.*, from Lindgren et al. [17 and 24].

Runnemalm and Hyun [18], Figure 7-6, combined this with error measures to create an automatic, adaptive mesh. They showed that it is necessary to account both for thermal and mechanical gradients in the error measure. The mesh created by the latter error measure is not so easy to foresee even for an experienced user. Therefore, the use of error measures is important.

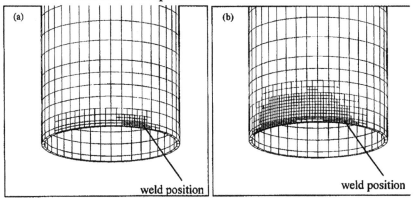

Figure 7-6: Adaptive meshing based on gradient in (a) thermal field and (b) gradients in thermal and mechanical fields, from [18 and 24]

The first three dimensional residual stress predictions of full welds appear to be by Lindgren and Karlsson [28], who used shell elements when modeling a thin-walled pipe. Karlsson and Josefson [27] modeled the same pipe using solid elements.

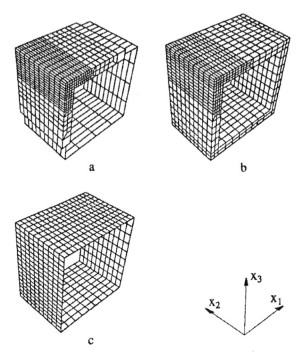

Figure 7-7: Finite element meshes used a) combined solid and shell element model b) solid element model c) shell element model, form [15].

The combination of shell and solid elements in the same model is of special interest in welding simulations. A thin-walled structure can be modeled with shell elements, but a more detailed resolution of the region near the weld will require solid elements. This was done by Gu and Goldak [14] for a thermal simulation of a weld and by Naesstroem et al. [15] for a thermomechanical model of a weld, Figure 7-7.

Shell elements can be successfully used in finite element calculations of thin walled structures. However, in the weld and the heat affected zone (*HAZ*) shell elements may not be sufficient, since the through thickness stress gradient is high in these regions. Naesstrom et al. [15] presented a combination of eight-nodes solid elements near the weld and four-nodes shell elements elsewhere. This combination of solid elements and shell elements reduces the number of degrees of freedom in the problem in comparison with the use of solid elements only. The residual stress results show that the combined model can be successfully used in structures where high stress and temperature gradients are localized to a narrow region i.e. a fine mesh in the heat affected zone and a coarse mesh of shell elements elsewhere.

McDill et al [16] developed a promising element formulation where a three-dimensional element with eight nodes and only displacements as model unknowns can be a solid or a shell element. It was demonstrated on a small-weld case in their paper. The element is based on the same element as in Lindgren et al [17] and Runnemalm and Hyun [18], making it possible to perform adaptive meshing with a combination of solid and shell elements. It will also be possible to determine adaptively whether an element should be treated as a solid or a shell.

Lindgren L.E.'s [24] recommendations say "the adaptive meshing and parallel computations are currently necessary to solve three-dimensional problems with the same accuracy as in existing two-dimensional models. The Eulerian approach is effective but less general".

Composite meshes are similar to adaptive meshes in that both are based on constraints, i.e., some nodes are declared to be linearly dependent on other nodes. Composite meshes differ from adaptive meshes in that fine elements need not be children of coarse elements, i.e., fine elements need not be formed by refining coarse elements.

Figure 7-4 shows an example of the weld pool being meshed with element boundaries on the liquid/metal interface. This weld pool mesh is parameterized so that weld pool shape can be defined dynamically during the analysis. We call this a parametric conforming weld pool mesh.

The weld pool dimensions for a double ellipsoid weld pool were:
- Front ellipsoid width, depth and length in meters are *(0.006, 0.000914 and 0.006)*.
- Rear ellipsoid width, depth and length are *(0.006, 0.000914 and 0.012)*.

In Figure 7-4 note that except very close to the weld pool, the temperature varies linearly through the thickness of the wall. This implies that almost everywhere only *2* nodes through the wall are needed to accurately capture the temperature variation. The five nodes that we used were not necessary and a significant reduction in computing time could have been achieved by using elements that were linear through the thickness of the wall and quadratic or cubic on the wall surfaces.

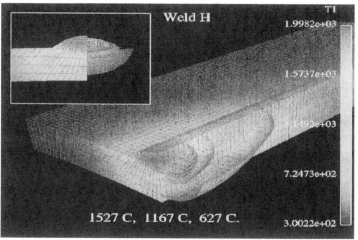

Figure 7-8: The temperature distribution in a composite mesh with fewer mesh parts

Figure 7-8 shows the temperature distribution in a composite mesh with fewer mesh parts. In this case there is no mesh part *WP1*. Thus this mesh makes no attempt to capture the weld pool geometry by placing element boundaries on the weld pool boundary. We call this a nonconforming weld pool mesh. In this case, elements near the weld pool are less deformed and there are fewer constraints between non-conforming meshes to evaluate.

Figure 7-9 shows displacement vectors near weld pool. Note that displacement vectors behind the weld pool are pulled backwards towards the start of the weld while displacement vectors ahead of the weld pool are pushed forward by the thermal wave advancing with the weld pool. It is the hysteresis of this cycle that leads to residual stress and distortion in welds.

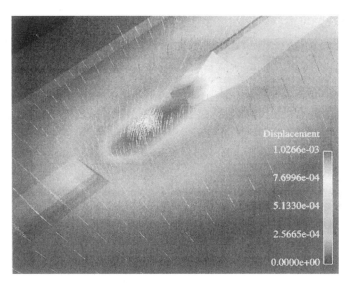

Figure 7-9: Displacement vectors are shown near the weld pool Discussion of progress in *CWM*

The structure is represented by a *FEM* mesh that ignores the details of the welded joints. The *FEM* mesh of the structure can be a separate mesh for each part or it can be single *FEM* mesh for the complete structure. If an automatic meshing capability is available, then it is convenient to define the geometry of the structure or parts by *CAD* files such as stereolithographic files.

In addition to the welded structure, a sequence of welds and weld passes for each weld must be specified. Such a sequence is a list of weld joints and for each weld joint a list of weld passes. Both lists are ordered in time. With each weld pass in each weld joint, a start time is specified.

Analyzing multipass welds as a series of single-pass welds is certainly the most rigorous, albeit costly, process. Multi pass welds have been analyzed by Ueda [33], Rybicki [30] and Leung [34]. To reduce the cost of separate analyses for each pass, several passes have often been lumped together in different ways, Figure 7-10. In some, only the last pass in a specific layer is analyzed [33]. In others, it would appear that the volume of the weld deposit of several passes or layers of passes are lumped together and the thermal history of a single pass located in the middle of the deposit is imposed during the stress analysis [30]. Another technique, lumps the thermal histories from several passes together, the temperature at any point, at any instant in time being the greatest value from any pass. All passes in a single layer, except the last, are lumped together. The last pass in any layer is treated separately [34]. Lumping layers together is inadvisable with the possible exception that the lumped layers remain a small proportion of the total thickness. For extremely large numbers of passes even these techniques may not be adequate.

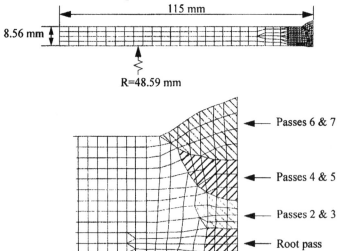

Figure 7-10: Finite element model used by Rybicki and Stonesifer [30], where seven weld passes were lumped into four passes, adopted from [24]

A somewhat more speculative approach was used by the Goldak et al [35] to analyze the resurfacing of a thin plate (hydroelectric turbine blade) that involved several hundred passes, figure 7-11. Each pass was 10 cm long; groups of 20 passes were done sequentially to cover 10 cm x 10 cm rectangular surface patches. The orientation of each patch, and the patch sequence were varied in order to minimize the deformation. The residual stress pattern of the multipass case was created by superimposing the residual stress pattern of each individual weld and taking the largest value from any individual pass. By building up the stress pattern for each patch and sequentially applying this pattern to the plate it was possible to obtain good qualitative agreement in both the deformed shape and optimum patch sequence. The danger in all these lumping techniques comes from the accuracy with which they model the sequencing effect.

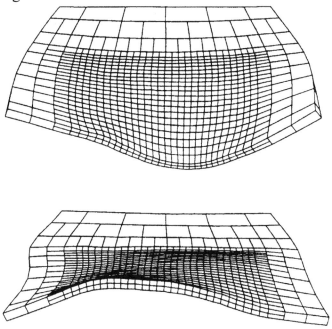

Figure 7-11: Deformed shapes of multipass, resurfaced, hydroelectric turbine blades; superimposed residual stress patterns. A: longitudinal patch sequence, B: transverse path sequence. The distortion is magnified *100* times.

Chakravarti and Goldak et al [35 and 36] developed a finite element model to predict the vertical distortions due to block welds in welding overlays on a flat plate. The finite element predictions are compared to experimental measurements. The *FEM* analysis uses a three stage model to do the transient thermal analysis, transient thermo-elasto-plastic, three dimensional analysis and elasto-plastic analysis of a plate containing six blocks using total strain from the blocks as initial strain in the plate. The results show reasonable agreement between the finite element predictions and experimental measurements. This gives confidence in the finite element model, and its potential to simulate more complex geometries where simple analytical expressions may be inadequate.

Lindgren's review [32] recommends that simplifying the multipass welding procedure by some kind of lumping technique must be exercised with care. All lumping and envelope techniques change the temperature history and will affect the transient and residual strains near the weld. Lumping by merging several weld passes that conserve the total heat input is preferred. Thus, the simulation will correspond to a multipass weld but with fewer weld passes than the original.

Figure 7-12 shows a structure to be welded as another example with several weld joints.

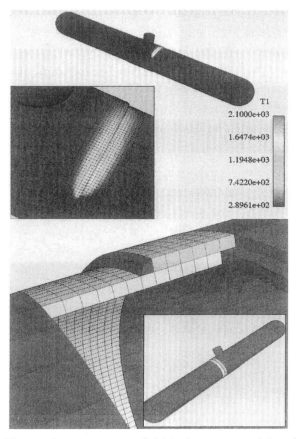

Figure 7-12: The transient temperature field is shown on one joint of the saddle in the upper picture. The mesh for this weld joint including filler metal is shown in the lower picture. For analyses this mesh is refined.

The *FEM* mesh for the structure is usually too coarse to be used for the analysis of a weld. In any case, it does not capture the geometry of the weld joint and the filler metal added in each weld pass. Therefore, a new *FEM* mesh is made for each weld joint. To do this, the specification of the weld procedure includes a *2D FEM* mesh. This *2D* mesh is parameterized by thickness of plates or other parameters associated with the weld joint. The parameterization of this *2D* mesh is such that it can be made to conform to the geometry of a cross-section of each weld joint. This *2D* mesh also includes the metal added for each pass in the weld joint as a separate part type for each weld pass.

For each weld joint this new weld joint mesh is composed with a subset of the *FEM* mesh of the structure to make the *FEM* mesh for the structure that will be the domain to be used for the *CWM* analysis of each weld pass in that joint. As each weld pass is made, the *FEM* elements in the filler metal added in the pass are added to the domain. This process is shown in Figure 7-13.

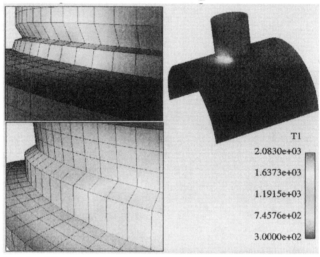

Figure 7-13: The filler metal added for the first and third pass of the Tee-joint weld joint are shown. The insets show the transient temperature field.

The next weld starts with the *FEM* mesh of the domain of the previous weld and adds the mesh of the next weld joint to a subset of the previous domain. Hence as the structure is fabricated, the mesh changes dynamically during each weld pass to capture the geometry and state of the structure including the geometry and state of welds that have been completed.

Figures 7-12 and 7-14 show transient thermal analyses of welds.

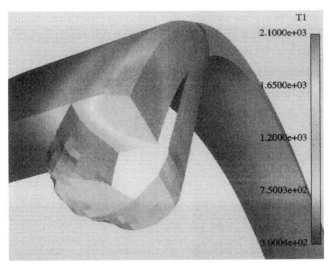

T1
2.1000e+03

1.6500e+03

1.2000e+03

7.5003e+02

3.0004e+02

Figure 7-14: The *1100°K* temperature isotherm of the weld in Figure 7-12

We learned two important lessons from our experience with parametric weld pool meshes. First and perhaps most important, the distorted elements in conforming weld pool meshes appear to have far larger discretization errors than the nonconforming weld pool mesh. This should not have been surprising because the effect of element distortion on accuracy has been studied and is reasonably well understood [11 and 12]. The second important lesson that we learned was the need to balance the overhead of working with composite meshes with the cost of solving. Although the cost of solving the nonconforming mesh is slightly higher, the cost of overhead is less and the total cost can be lower.

In pursuing our ultimate objective of *CWM* analyses of welded structures that have many, multipass welds; we have developed a software environment that separates the structural design, weld joint design and production welding stages. Structural design specifies the parts to be welded; in particular their geometry, their material types and any relevant internal variables and boundary conditions needed to constrain rigid body motion. Weld joint design specifies a curvilinear coordinate system for each weld joint. Each weld pass on each weld joint has a start point, end point and a start time. Production welding specifies the welding procedure for each weld joint. The welding procedure defines the welding parameters for each pass, e.g., current, voltage, speed, weld pool size, shape and position in a cross-section of the weld joint. These three data sets fully define the process of welding a structure and contain the data needed to perform a computational welding mechanics analysis for fabrication of a complete structure.

7.5 Real-Time for *CWM*

Is real-time Computational Welding Mechanics feasible? For the question to have meaning, it is necessary to first specify what we mean by a solution and what errors in the solution are tolerable. We define a solution to be the transient temperature, displacement, strain and stress and microstructure evolution evaluated at each point in space and time in a weld. In the authors' opinion, numerical errors in stress and strain in the range *(5% and 10%)* are acceptable.

By numerical errors we mean the difference between the numerical solution and an exact solution to the mathematical problem being solved. By real errors we mean the difference between the numerical solution and experimentally measured values in which there are no experimental errors. It is difficult to reduce real errors below *1%* because of errors in available data for material properties such as thermal conductivity and Young's modulus. Numerical errors in the solution that are much larger than *10%* often would not be accepted. The numerical error estimates do not include errors in material properties or in microstructure evolution. We assume that errors in geometry will be held to less than the maximum of *1%* or *0.5 mm* except for some details.

The uncertainties of the computerized simulation of the cross-sectional geometric parameters of welds are investigated by Sudnik et al [25], using laser beam welds as an example. It has been discussed how to estimate the different errors in the simulations, that are modelling, parametrical (e.g., uncertainty in material and process data) and numerical errors.

The error propagation rule according to Gauss together with error sensitivity coefficients is used as the basis. The uncertainties of simulation are formally dealt with in the same manner as it is usual with the uncertainties of testing results. The simulation error is considered as being composed of modeling errors, parametrical errors and numerical errors. Simulation error and testing error together result in the verification error or prediction confidence. The example comprises a CO_2 laser beam welds in steel simulated by the computer program *DB-LASIM* resulting in a modeling error of about *10%* and a prediction error of about *13%* (standard deviations).

The effect of arbitrarily chosen extreme variations of the material properties on the cross-sectional simulation results is additionally visualized by Figure 7-15.

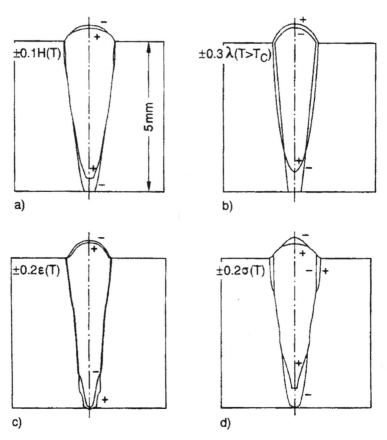

Figure 7-15: Effect of arbitrarily chosen extreme uncertainties of material properties on the cross-sectional simulation result, i.e. of $\pm 10\%$ enthalpy (a), of $\pm 30\%$ thermal conductivity at $T > T_c = 769°C$ (b), of $\pm 20\%$ absorption coefficient (c) and of $\pm 20\%$ surface tension, adopted from Sudnik et al. [25].

There is a varying need for accuracy during different stages of the design process. Isaksson and Runnemalm developed a systematic approach based on simplified simulations used to create a weld response matrix *(WRM)* used in the preliminary design stage of a welding procedure. The *WRM* should relate changes in design variables with changes in variables that are important for the desired performance.

7.5.1 Current Performance for CWM

We begin by considering thermal-microstructure evolution of an arc weld that is a *100 mm* long and made with a welding speed of *1.785 mm/s* and is completed in *71.4 s*. Of course, many production welds use much faster welding speeds. We were (1993) able to do a high resolution *3D* transient thermal analysis in only a few seconds. This analysis involved *40* time steps and *8,718* 8-node bricks. The mathematics used is described in details in [9] and also in chapter *V*.

To a large extent, the time domain for the thermal-microstructure analysis in many welds can be decomposed into three stages: starting transient, steady state and stopping transient. If this were done, it would be feasible to achieve real-time analysis today.

There are many reasons why a thermal stress analysis of welds is a much more challenging problem than a thermal-microstructure analysis. The thermal-microstructure analysis only involves material a short distance from the weld path, usually less than ten weld pool diameters. Only this relatively small region near the weld need be analyzed for temperatures and microstructure evolution. Usually, this width is less than *10 cm*. In contrast, in a thermal stress analysis the complete structure being welded is in quasi-static equilibrium. Thus thermal stresses generated by the welding process can travel over the complete structure. This makes it much more difficult to do the analysis in a relatively small region around the weld. In particular, it is difficult to choose realistic boundary conditions for a small region around the weld.

Another factor is that the mesh used for thermal-stress analysis must be finer than the mesh used for thermal analysis. For the examples described above we use an 8-node brick for stress and treat the temperature in the element as piece-wise constant. The reason for this is that strain is the gradient of the displacement. The gradient operator essentially reduces the order of the strain field to one less than the order of the displacement field. If the thermal strain is to be consistent with the strain from the displacement gradient, the thermal element should be one order lower than the displacement element.

Another reason that thermal stress is more challenging than thermal analysis is that thermal analysis is strictly positive definite because of the capacitance matrix. On the other hand, the quasi-static thermal stress analysis is only made positive definite by constraining rigid body modes. In addition, a temperature increment in the order of *100 °K* generates a thermal strain equivalent to the yield strain. While larger temperature increments do not cause serious difficulties for the thermal solver, temperature increments that generate stress increments larger than the yield strength make it difficult to do an accurate thermal stress analysis.

If one used a *20*-node brick with *60* displacement dofs for thermal stress analysis and an *8*-node brick with *8* temperature dofs for thermal analysis, the thermal stress analysis would have *60/8=7.5* times more equations to solve. For a regular mesh topology, each global equation has *243* nonzero terms compared to *27* nonzero terms for the thermal solver. Thus a very rough lower bound estimate is that a thermal stress solver is *(60/8) · (243/27) = 67.5* times more expensive than a thermal solver. It also requires more than *70* times as much memory because stress and internal variables must be stored at Gauss points. These rough estimates agree with our experience that the thermal stress analysis in *CWM* is roughly *10* times more expensive than the thermal-microstructure analysis when *8*-node bricks are used for both analyses. We would expect the stress analysis to be roughly *100* times more expensive if *20*-node bricks were used for stress analysis and *8*-node bricks for thermal analysis.

The cost of *CWM* is roughly linearly proportional to the number of elements in the mesh, the number of time steps, the number of nonlinear iterations per time step and the time required for each nonlinear iteration. There are opportunities to optimize the mesh and reduce the number of elements. In particular, the use of shell elements could reduce the number of DOFs in a problem. A shell element usually has five DOFs at a node compared to a brick that has six.. However, near the weld solutions are truly *3D* and shell elements could introduce large errors. We have long favored the use of local *3D* transient analysis near the weld pool and shell elements farther from the weld pool where the assumptions of shell theory are valid. There are also opportunities to take longer time steps. A steady state Eulerian analysis of an infinitely long weld in a prismatic geometry would only require one solution step. This would require approximately a few seconds of *CPU* time for thermal analysis and a few minutes for a thermal stress analysis. This could easily be done in real time for a sufficiently long weld but would not capture the starting and stopping transients.

As of July 2004, Goldak Technologies Inc. is able to do a thermal stress analysis of a weld in a complex structure at a speed of 0.5 mm/s or 2.0 m/hr using a single CPU 3.2 GHz Pentium. By the end of 2004, their goal is to increase the computing speed into the range of 2.5 to 5 mm/s or 10 to 20 m/hr. This would be real-time CWM for many, though not all, industrial arc welding processes.

7.5.2 Implications of real-time CWM

If our prediction that *CWM* can be done in real-time is correct then *CWM* is likely to be used routinely in industry. This raises the question of what will be the impact of routine use of *CWM* in industry. We believe this will dramatically change both engineering and research in the three main components of welding technology for all types of welded structures; materials engineering, structural engineering and weld process development, [7].

Materials engineering will be able to simulate microstructure evolution much more accurately. While research on the evolution of microstructures is well advanced, research on the difficult issue of predicting or estimating material properties for a given microstructure with particular emphasis on failure mechanisms is just beginning. The other hard materials research issues will involve local bifurcations such as nucleation of phases, shear bands and porosity; (what materials engineers call nucleation, mathematicians often call bifurcations). These are the three fundamental bifurcations in material engineering. Shear bands involve only deviatoric stress and strain. Porosity involves only volumetric stress and strain. The computational mechanics of these bifurcations is a fairly hard research issue that has been almost totally ignored in the welding literature. It will become a major research topic in materials engineering for welding.

Structural engineering for welded structures will focus on the life cycle of welded structures from conceptual design, to manufacture, in-service use, maintenance and decommissioning. While life cycle engineering of welded structures is not novel, *CWM* is usually ignored. For example, while the role of residual stresses in welded structures has long been recognized, it has seldom been included in structural analysis. We believe that *CWM* will enable the effects of residual stress and weld microstructure to be integrated into the life-cycle engineering process. More research will focus on buckling of welded structures. The hardest *CWM* research issues will be local failure modes due to shear band or porosity formation, i.e., ductile fracture. We expect *CWM* to become an integral part of structural engineering of welded structures.

The third component of welding technology will be weld process development. The focus will be on the weld pool, the arc and laser physics. This involves the hardest research issue of all - turbulence. Because turbulence is a chaotic phenomenon, *CWM* of weld processes is likely to strive to resolve essentially small process variations and short time behavior.

Perhaps the most important change arising from the routine use of *CWM* in industry is that *CWM* will become closely tied to experiments, including experiments on real production structures. In the past, it has been too expensive to do experiments to validate *CWM*. If *CWM* is used by industry, in a sense the experiments become free. This will lead to tight coupling between experimental data and *CWM* analysis. This in turn will allow both experiments and *CWM* to be highly optimized and validated. When this stage is reached, *CWM* will have become a mature technology and welding technology will have a much stronger science base.

References

1. Yang Y.P. and Brust F.W. Welding-induced distortion control techniques in heavy industries, Symposium on weld residual stresses and fracture 2000, ASME Pressure Vessels and Piping Conference, Seattle WA USA, July 23-27 2000
2. The procedure Handbook, Lincoln Electric Company, 12th Ed., 1973
3. Henwood C. Bibby M.J., Goldak J.A. and Watt D.F. Coupled transient heat transfer microstructure weld computations, Acta Metal, Vol. 36, No. 11, pp 3037-3046, 1988
4. Goldak J.A., Breiguine V., Dai N., Zhou J. Thermal stress analysis in welds for hot cracking, Editor H. Cerjak H., 3rd Seminar, Numerical Analysis of Weld ability, Graz, Austria, Sept. 25-26 1995.
5. Goldak J.A., Mocanita M., Aldea V., Zhou J., Downey D., Dorling D. Predicting burn-through when welding on pressurized pipelines, Proceedings of PVP'2000, 2000 ASME Pressure Vessels and Piping Conference, Seattle Washington USA, July 23-27 2000
6. Goldak J. and Mocanita M. Computational weld mechanics and welded structures,
7. Goldak J.A., Mocanita M., Aldea V., Zhou J., Downey D. and Zypchen A. Is real time CWM feasible? Recent progress in CWM; 5th International Seminar Numerical Analysis of Weldability, IIW Com. IX, Graz-Seggau Austria, Oct. 1999
8. Goldak J. Input data for computational weld mechanics, Carleton University, Nov. 1 2002
9. Goldak J.A., Breiguine V., Dai N., Hughes E. and Zhou J. Thermal Stress Analysis in Solids Near the Liquid Region in Welds. Mathematical Modeling of Weld Phenomena, 3 Ed. By Cerjak H., The Institute of Materials, pp 543-570, 1997
10. Sudnik V.A. Research into fusion welding technologies based on physical-mathematical models, Welding and Cutting, pp 216-217, 1991
11. Zienkiewicz O.C. and Taylor R.L. The finite element method, Fourth Edition, Vol. 1, page 170, McGraw-Hill, 1988
12. Wachpress E. High order curved finite elements, Int. J. Num. Meth. Eng. 17, pp 735-745, 1981
13. McDill JM, Oddy AS. and Goldak JA. An adaptive mesh-management algorithm for three-dimensional automatic finite element analysis, Transactions of CSME, Vol. 15, No 1, pp 57-69, 1991
14. Gu M. and Goldak J.A. Mixing thermal shell and brick elements in fea of welds, Proc. of 10th Int. Conf. on Offshore Mechanics and Arctic Eng. (OMAE), Vol. III-A Materials Eng., p 1, 1991
15. Naesstroem M. Wikander L., Karlsson L., Lindgren L-E. and Goldak J. Combined 3D and shell modeling of welding, IUTAM Symposium on the Mechanical Effects of Welding, p 197, 1992

16. McDill JMJ., Runnemalm KH and Oddy AS. An 8- to 16- node solid graded shell element for far-field applications in 3D thermal-mechanical fea, the 12th Int. Conf. on Mathematical and Computer Modeling and Scientific Computing, Chicago USA, 2-4 Aug. 1999

17. Lindgren L-E., Haeggblad H-A., McDill JMJ. and Oddy AS. Automatic remeshing for three-dimensional finite element simulation of welding, Computer Methods in Applied Mechanics and Engineering, Vol. 147, pp 401-409, 1997

18. Runnemalm H. and Hyun S. Three-dimensional welding analysis using an adaptive mesh scheme, Computer Methods in Applied Mechanics and Engineering, Vol. 189, pp 515-523, 2000

19. Siva Parsad N. and Sankaranarayanan T.K. Estimation of residual stresses in weldments using adaptive grids, Computers & Structures, Vol. 60, No.6, pp 1037-1045, 1996

20. McDill JMJ. and Oddy AS. A non-conforming eight to 26-node hexahedron for three-dimensional thermal-elastoplastic finite element analysis, Computers & Structures, Vol. 54, pp 183-189, 1995.

21. McDill JMJ, Goldak JA, Oddy AS, and Bibby MJ. Isoparametric quadrilaterals and hexahedrons for mesh-grading elements, Comm. Applied Nu. Methods, Vol. 3, pp 155-163, 1987

22. McDill JM, Oddy AS. and Goldak JA. An adaptive mesh-management algorithm for three-dimensional automatic finite element analysis, Transactions of CSME, Vol. 15, No 1, pp 57-69, 1991

23. McDill JMJ. and Oddy AS. Arbitrary coarsening for adaptive mesh management in automatic finite element analysis, J Math. Modeling & Sci. Comp., Vol. 2B, pp 1072-1077, 1993

24. Lindgren L-E. Finite element modeling and simulation of Welding Part III, Efficiency and integration, J of Thermal Stresses 24, pp 305-334, 2001

25. Sudnik W., Radaj D. and Erofeew W. Validation of computerized simulation of welding processes, Ed. Cerjak H., Mathematical Modeling of Weld Phenomena, Vol. 4, pp 477-493, 1998

26. Sudnik W., Radaj D., Breitschwerdt S. and Erofeew W. Numerical simulation of weld pool geometry in laser beam welding; J Phys. D. Appl. Phys. 33, pp 662-671, 2000

27. Karlsson R.I. and Josefson B.L. Three dimensional finite element analysis of temperature and stresses in single pass butt welding, J Pressure Vessel Technology, Trans. ASME, 1987

28. Lindgren L.E. and Karlsson L. Deformation and stresses in welding of shell structures, Int. J. for Numerical Methods in Engineering, Vol. 25, pp 635-655, 1988

29. Goldak J. Keynote address modeling thermal stresses and distortions in welds, Resent Trends in Welding Science and Technology TWR,89, Proc. of the 2nd International Conference on Trends in welding Research, Gatlinburg Tennessee USA, 14-18 May 1989

30. Rybicki E.F. and Stonesifer R.B. Computation of residual stresses due to multipass welds in piping systems, ASME J., Pressure Vessel Technology, Vol. 101, pp 149-154, 1979
31. Radaj D. Heat effects of welding: temperature field, residual stress, distortion. Springer, 1992
32. Lindgren L-E. Finite element modeling and simulation of welding Part I Increased complexity, J of Thermal Stresses 24, pp 141-192, 2001
33. Ueda Y. and Murakawa H. Applications of computer and numerical analysis techniques in welding research, Trans. of JWRI, Vol. 13, No. 2, pp 337-346, 1984
34. Leung C.K. and Pick R.J. Finite element analysis of multipass welds in piping systems, Trans. ASME, J. Press. V. Tech., Vol. 101, pp 149-154, May 1979
35. Goldak J. Oddy A., Gu M., Ma W., Mashaie A. and Hughes E. Coupling heat transfer, microstructure evolution and thermal stress analysis in weld mechanics, IUTAM sym. Lulea Sweden, June 1991
36. Chakravarti A.P., Malik L.M., Rao A.S. and Goldak J.A. Prediction of distortion in overlayed repair welds, 5th International Conference on Numerical Methods in Thermal Problems, Swansea UK, July 1985

Chapter VIII

Fracture Mechanics

8.1 Introduction and Synopsis

The development and production of High Strength Low Alloy *(HSLA)* linepipe steels has been growing rapidly because of their desirable combination of high strength, low temperature toughness, weldablity and low cost.

With the advent of the welding as the major method of fabrication, cracking in the heat affected zone has become a serious problem particularly in large and continuous structures.

One of the major problems in the welding of steels has been the type of cracking generally known as hydrogen induced cold cracking, which is described in chapter *VI*.

At high carbon levels, the heat affected zone cracking was severe because of the formation of brittle martensite as a result of rapid cooling in welding. In the presence of hydrogen gross cracking was inevitable. Recognizing the deleterious effect on notch toughness and weldability by carbon in welding, the trend towards newly improved materials has resulted in the development of lower carbon steels. Quenched and tempered steels were produced using alloy elements such as manganese and silicon as the main strengtheners in solid solution. These additives increased the hardenability of the steels to make heat treatments possible. With the concurrent development of low hydrogen electrodes, these steels were easily welded without any cracking problem at least in thin sections. In general, preheat is only required in thicker section and when the restraint is high. However, since mechanical properties of these steels were strictly controlled by careful heat treatments during processing, the thermal cycle in welding can significantly impair strength and toughness in the fusion zone and the heat affected zone. This limitation in the use of quenched and tempered steels has paved the way for developing lower cost, high strength and toughness low alloy steels. There are five major factors determining the strength, impact toughness and weldability of low alloy steels. They are carbon content, dislocation density, precipitation hardening, solid solution hardening and grain size. Only reduction of grain size significantly improves both strength and toughness; the other strengthening mechanisms improve strength at the expense of toughness, [5 and 11].

Weld solidification cracking (hot cracking) has been a persistent problem in a variety of engineering alloys. The formation of solidification cracks result from the combined effects of metallurgical and mechanical factors. The metallurgical factors relate to conditions of solidification, grain size, presence of low-melting phases, etc. The mechanical factors relate to conditions of stress/strain/strain rate developed near the trailing edge of a weld pool during solidification. In other words, solidification cracking is a result of the competition between the material resistance to cracking and the mechanical driving force. The material resistance to cracking is primarily influenced by alloy composition, the welding process and heat input. The mechanical driving force depends upon the welding process, heat input, joint configuration and rigidity and thermo mechanical properties of alloys.

Although the precise mechanisms responsible for solidification cracking and the remedies for its prevention are still not clear, research efforts have revealed a more or less qualitative picture about the nature of such cracking. A hot crack requires strain localization. Therefore it must require strain softening. It appears that rate-dependent, i.e. viscous, behavior plays a critical role in hot cracking. The solidification of the weld pool is a non-equilibrium process and depending on growth speed and temperature gradient and can result in the formation of columnar microstructure or dendritic microstructure under the condition of constitutional super-cooling.

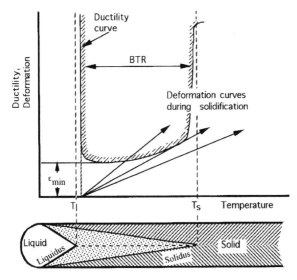

Figure 8-1: The material resistance versus mechanical driving force concept, adopted from [7].

Hot cracking of welds is an important problem in the welding industry. It is thought to occur in a temperature range that begins at the start of solidification and ends soon after solidification is entirely complete.

During the later stage of solidification, there exists a brittle temperature range *(BTR)* in which the strength and ductility of the alloy are very low as some low melting-point constituents segregate between dendrites and form liquid films. At the same time, stresses/strains arise from solidification shrinkage, thermal contraction of parent metal and external restrains.

Like many other cracking problems, solidification cracking occurs when the mechanical driving force exceeds the material resistance to cracking. This concept is illustrated by Feng et al [7] in Figure 8-1.

Hot cracking is observed to be sensitive to chemical composition. It is thought that a critical strain and critical strain rate exists. The critical strain for hot cracking vs. temperature at constant strain rate for some *0.2%C* steel could be described by the points *(0.1, 1490)*,

(0.04, 1480), (0.1, 1470), see 5-6-2. As the strain rate is decreased, the critical strain decreases but below a critical strain rate, hot cracking does not occur, [3].

Hot cracking has been studied in depth from the viewpoint of the effect of metallurgical variables. In addition a number of experimental techniques have been developed to measure sensitivity to hot cracking and study the phenomenon of hot cracking. These include the Houldcroft, Varestraint and Sigmajig tests. The Houldcroft test varies the restraint by sawing slits into the plate in an attempt to vary the strain and strain rate. No external forces are applied. The Sigmajig test which is commonly used for sheet metal applies a transverse force or stress. The Varestraint test applies a displacement that strains the weld pool region, [3]. In the *1980s,* Matsuda et al [16] published a series of basic research results on weld metal solidification cracking phenomenon based on a new technique, Measurement by means of In-Situ Observation *(MISO).*With assistance of the *MISO* technique, Matsuda was able to directly measure the ductility around the solidification crack tip on a very local scale (about *1mm* in gauge length). Recently, Lin and co-worker [17] improved the measurement of the *BTR* and demonstrated that, by using the new measurement methodology developed, the *BTR* is material specific and independent of the testing conditions.

In plain carbon steel structures, it has long been recognized that brittle fractures almost never propagate along the heat affected zone *(HAZ)* of welds; consequently little effort was made to measure the toughness of the *HAZ*. However, with the increasing use of medium and high strength steel and the development of high heat welding processes, the toughness of the *HAZ* was questioned and a number of papers were published on the subject.

The Charpy *V*-notch test is a standard laboratory test for welded joints. Goldak and Nguyen [5] present data proving that conventional Charpy-V notch *(CVN)* testing of narrow zones in electron beam *(EB)* welds can produce dangerously misleading results and describe a simple economical technique for measuring the true ductile brittle transition temperature of narrow zones in welds. They suggest the Cross Weld Charpy Test *(CWCT),* which actually does measure *CVN FATT* (Fracture Appearance Transition Temperature) in *EB* weld metal. In addition to *CVN* tests in which the weld plane is parallel to the fracture plane, Figure 8-2, the *CWCT* uses a conventional Charpy-*V* notch specimen except that the weld plane is perpendicular to the fracture plane to force the fracture to pass through all zones of the weld.

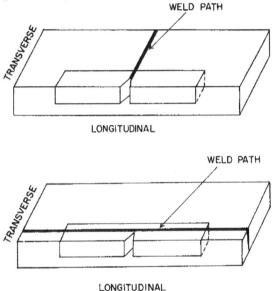

Figure 8-2: Weld position in: top-conventional Charpy *V*-notch specimen; bottom-cross weld Charpy *V*-notch specimen

The competition between brittle and ductile modes of failure in a weld specimen under dynamic loading has been analyzed by Needleman and Tvergaard [20], in terms of a micromechanically based material model. In addition the ductile-brittle transition for different weld joints has also been investigated by plane strain numerical analyses of Charpy impact specimens by the same authors [4].

In the *1970s*, Chihoski published three papers [2, 14 and 15] on an experimental study of displacements near the weld pool. His objective was to understand the strain field around a weld pool and the effect of welding variables, such as welding speed, on the strain field and the susceptibility to hot cracking. He used a Moire' Fringe method. He studied three edge welds and three bead-on-plate welds. These were *TIG* welds in thin aluminum-copper alloy, *Al 2024*, plates. He used welding speeds of *6, 13* and *20 ipm*. He found a compressive region near the weld pool. Roughly the thermal expansion near the weld pool causes a compression region that is later overwhelmed by the shrinkage stresses as the weld cools, Chihoski observed that the location of this compression region relative to the weld pool was a function of welding speed. By placing this compression region in the region sensitive to hot cracking, he was able to weld *Al 2024*, an alloy sensitive to hot cracking, without hot cracking.

In addition to the study of metallurgical affects and microstructure, which are clearly important to hot cracking in welds, it is also necessary to develop and test a capability for stress analysis near the weld pool. Chapter *V* presents a preliminary attempt to develop a capability to do quantitative analysis of the stress and strain near the weld pool. In addition experimentally determined constitutive equations or other forms of experiments such as those of [14 and 19] will be needed to validate the analysis. Ultimately microstructure models of dendrites and interdendritic liquids will be needed. Even with these limitations, the model introduced in chapter *V* appears to be a useful step towards quantitative analysis of stress and strain in the region susceptible to hot cracking in welds.

Feng and co-worker [7] present also the development of a finite element analysis procedure and the calculated dynamic stress/strain evolution that contributes to the formation of solidification cracks in the cracking susceptible temperature range.

As interest in enlarging the gas throughput increases, the use of larger diameter and higher pressure gas transmission pipelines will rise. There will then be an increasing need for reliable pipeline design, inspection and maintenance procedures that will minimize service failures. Concern for the possibility of ductile fracture propagation in gas transmission pipelines stems from two main sources. With the ever-expanding gas transmission system, the probability rises of a third party inflicting damage severe enough to initiate fracture. Further, as the current systems age, the probability of insidious corrosion damage growth producing a local rupture event increases. Failure through either of these mechanisms can result in the initiation of a long-running ductile fracture. A methodology has been developed by O'Donoghue et al [6] that can be used on a routine basis to predict the possibility of the occurrence of long-running cracks in gas transmission pipelines.

8.2 Review and Model Development

Since the *1950s*, more than one hundred separate and distinct weldability tests aiming at assessing the solidification and liquation cracking susceptibility have been devised and completely new or modified tests are under continued development. However, the design and fabrication of solidification crack-free structures has not been completely successful despite tremendous effort. A critical problem remains the lack of adequate techniques to quantify the stress/strain variations during the solidification process. There are concerns in appropriately quantifying laboratory weldability testing results. This can be seen by comparing the results of Arata et al [16 and 18] in which the materials were tested under similar welding conditions. It was found that the ductilities in the *BTR* measured by the *MISO* technique were often an order of magnitude higher than those measured by the augmented strain of the Trans-Varestraint test, whereas the *BTR* itself generally showed little change. In fact, the minimum ductilities obtained by the *MISO* technique were often quite measurable for alloys that are generally regarded as being highly susceptible to weld solidification cracking, [7].

More importantly, it is extremely difficult to reliably apply laboratory weldability testing results to the actual fabrication problems, since the mechanical driving force under the actual fabrication conditions has rarely been quantitatively determined.

Studies of hot cracking require more accurate models. It is important to have a correct description of the material behavior in order to have an accurate model. The more important mechanical properties are Young's modulus, thermal dilatation and parameters for the plastic behavior. The influence of these properties at higher temperatures is less pronounced on the residual stress fields. The

material is soft and the thermal strains cause plastic strains even if
the structure is only restrained a little as the surrounding, initially
cold material acts as a restraint on the heat-affected-zone. If a cut-off
temperature is used, the strain and strain rate are expected to be
under-estimated and stress over-estimated.

According to the Feng's et al [7] study, direct observation of the
solidification cracking of Type *316* stainless steel in the Sigmajig
test has revealed that the weld centreline cracking was often initiated
at a location some distance behind the apparent trailing edge of the
weld pool, i.e., at a temperature below the bulk solidus of *316* steel
(1645 °K). This study assumes that depending upon the testing
conditions, cracking would initiate anywhere between *1600 °K* and
1300 °K as long as the stress/strain condition and the microstructure
at that particular temperature favoured such an event.

In the Sigmajig test, a specimen is pre-stressed by a pair of steel
bolts before welding. The pre-stress is maintained by two stacks of
Bellville washers in the load train, which have a displacement/load
curve with a slope of *6.167×10⁻⁴ mm/N* for each stack of washers,
Figure 8-3:

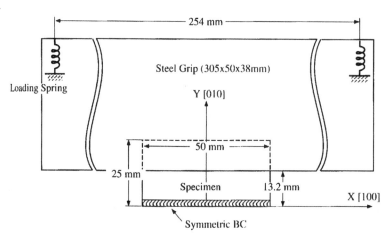

Figure 8-3: Geometry representation of the Sigmajig test and boundary conditions
in the mechanical model used by Feng et al [7].

Feng et al [7] have discussed development of finite element
models that included the solidification effect in the weld pool. The

models are used to calculate the local stress/strain evolution in the solidification temperature range in which weld solidification cracking takes place. The results presented in his study indicate the possibility of using the finite element analysis, when properly formulated, to capture the local stress/strain conditions in the vicinity of a moving weld pool. The models not only show very good agreement with the quantitative measurements of deformation patterns in the vicinity of the weld pool in aluminium alloys, but also correlate very well with the experimental observations of cracking initiation behaviour of a nickel-based super-alloy under various welding and loading conditions during the Sigmajig weldability test. Figure 8-4 shows the welds with cracks in some selected specimens.

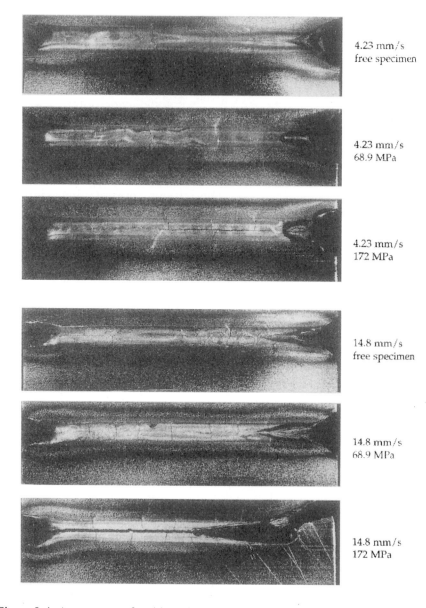

4.23 mm/s
free specimen

4.23 mm/s
68.9 MPa

4.23 mm/s
172 MPa

14.8 mm/s
free specimen

14.8 mm/s
68.9 MPa

14.8 mm/s
172 MPa

Figure 8-4: Appearance of welds and solidification cracks. Welding direction was from left to right. From top to bottom: *4.23 mm/s* free; *4.23 mm/s 68.9 MPa*; *4.23 mm/s 172 MPa*; *14.8 mm/s* free; *14.8 mm/s 68.9 MPa*; *14.8 mm/s 172 MPa*, adopted from [7].

Experiment has revealed that the centreline solidification cracking, if it occurs, would initiate at the starting edge of the specimen. Centreline cracks were observed at the high and low loading conditions, while the medium loading condition did not cause any centreline crack, Figure 8-4.

Intuitively, it is difficult to comprehend why centerline cracking occurs in the stress-free specimen but not in a moderately pre-stressed specimen based on the threshold pre-stress concept. However, this phenomenon can be readily interpreted in terms of the local stress evolution at the crack initiation site located in the starting edge of the weld. Figure 8-5 plots the transverse stress evolution at the crack initiation site only. Compressive transverse stresses initially develop upon solidification from *1600 °K*, then changing to tension before the temperature drops to *1500 °K* for all three pre-stress conditions. However, the tensile stress in the *68.9 MP*a case is always lower than the other two cases over the entire crack initiation temperature range, [7].

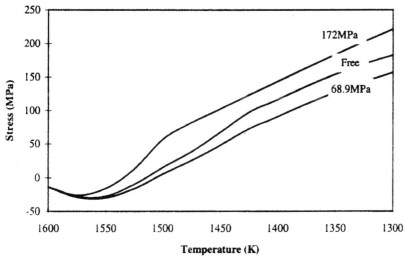

Figure 8-5: Transverse stress evolution responsible for centerline cracking initiation at the starting edge of the specimen, adopted from [7].

According to Figure 5-24, the compressive stress region is larger and tends to shift towards the rear of the weld pool for a faster weld. This is the effect observed by Chihoski and suggests the faster weld would be less susceptible to hot cracking because more of the temperature region susceptible to hot cracking is in compression. Data of this type would be also useful to compare with Sigmajig test data.

Dangerously misleading results obtained by conventional Charpy *V*-notch *(CVN)* testing of narrow zones in electron beam welds were reported by Goldak and Nguyen [5 and 11]. The fracture appearance of the various weld zones as measured with a binocular microscope and a scanning electron microscopy *SEM* is showed in Figure 8-6.

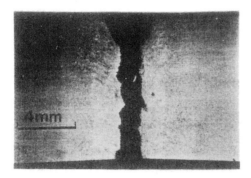

Figure 8-6: Charpy impact fracture paths for the notch: a) at the first notch position *-30°C* or *-22°F* and *5.4 joules* (fusion zone) ; b) at the second notch position *-31°C* or *-24°F* and *4 joules*; c) at the third notch position *-19°C* or *-2°F* and *68 joules* (grain coarsened *HAZ*)

A careful study of the fracture path revealed the disturbing fact that no tough fractures were ever observed in the weld metal for this weld at any temperature, Figure 8-7. The absorbed energy versus temperature curve for the Charpy *V*-notch test is often used to characterize the ductile-brittle transition in steels.

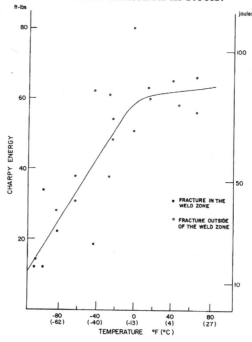

Figure 8-7: *CVN* test results showing fracture traveled in base metal above *-40 °C (-40°F)* and in the weld metal below *-80°C (-112°F)*

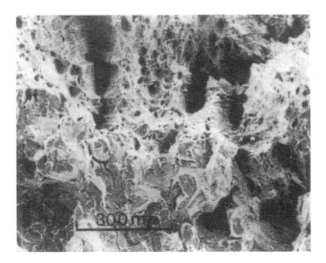

Figure 8-8: *SEM* fractograph of cross weld Charpy test specimen *(6 kJ/in.* or *0.24 kJ/mm, -3 °C* or *+17.6 °F* and *46 ft-lb* or *62.4 J)* showing cleavage facets indicative of brittle fracture on the weld in the lower half of the photograph; the base metal has failed ductilely.

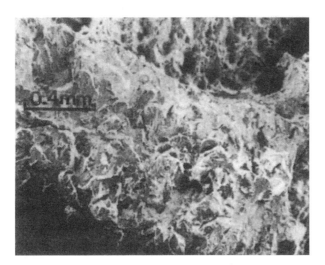

Figure 8-9: *SEM* fractograph of cross weld Charpy test specimen *(6kJ/in.* or *0.24kJ/mm, +21 °C* or *+70°F* and *48 ft-lb* or *65.1 J).* The fracture in the weld metal on the lower half of the figure is entirely brittle; the base metal has failed ductilely by void coalescence

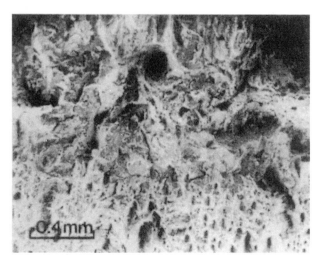

Figure 8-10: *SEM* fractograph of cross weld Charpy test specimen *(6kJ/in. or 0.24kJ/mm, +50 °C or +122°F and 50 ft-lb or 67.8 J)*. The fracture in the weld in the center of the photograph shows both cleavage facets and ductile dimples; the lower third of the photograph shows a ductile fracture in the base metal.

Above the *CVN (FATT)* all fractures occurred in base metal by a plastic hinge mechanism. Below the *CVN FATT* all fractures were brittle and traveled in the weld metal. At *-3* and *21 °C (27 and 70 °F)* the *CWCT* results shown in Figures 8-8 and 8-9 illustrate a completely brittle fracture in the fusion zone comprised entirely of cleavage facets.   At *+50 °C* or *122 °F*, Figure 8-10 shows equal areas of cleavage fracture and ductile dimpling in the fusion zone indicating a *"FATT"* for the fusion zone in the *CWCT*.

Tvergaard and Needleman [21 and 22] have applied a micromechanically based material model in a plane strain analyses of the Charpy *V*-notch test and have found that the experimentally observed behavior is well reproduced by the computations. The ductile-brittle transition for a weld is investigated by numerical analysis of Charpy impact specimens by Tvergaard and Needleman [4]. In this study, plane strain analyses of the Charpy *V*-notch test are carried out for different welded joints, to investigate the ductile-brittle transition in different parts of the weld, Figure 8-11. The *HAZ* is of thickness w_2 and the weld material lies between the two *HAZ* regions. Attention is focused on two cases. In one case, $x_c^2 = 0$ giving a weld that is symmetrical about the notch, while in the other case, $x_c^2 = 6.5$ *mm*, so that one *HAZ* is located directly in front of the notch.

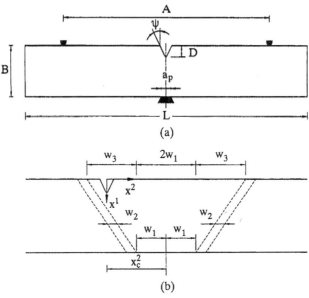

(a)

(b)

Figure 8-11: A plane strain model of the standard Charpy *V*-notch specimen analyzed by Tvergaard and Needleman [4] with the geometry of the Charpy specimen (a): The dimensions in the plane of deformation being *L=55 mm, B=10 mm, D=2 mm, R=0.25 mm, A=40 mm* and $\psi=22.5°$. The geometry of the weld (b): $w_1 = 0.5$ *mm*, $w_2 = 0.5$ *mm* and $w_3 = 8$ *mm*.

In relation to the procedures prescribed in European Standards, the planar analyses can directly represent impact test specimens with the notched face parallel to the surface of the test piece, while specimens with the notched face perpendicular to the surface of the test piece would generally require a full three dimensional analysis. The micromechanically based material model is used to represent the elastic-viscoplastic properties and the failure behavior in the base material, the weld material and the *HAZ* and specimens with a number of different notch locations in and around the weld are analyzed.

A convected coordinate Lagrangian formulation is used with the dynamic principle of virtual work written as, adopted from [4]:

$$\int_V \tau^{ij} \delta E_{ij} dV = \int_S T^i \delta u_i dS - \int_V \rho \frac{\partial^2 u^i}{\partial t^2} \delta u_i dV \tag{8-1}$$

with:

$$T^i = (\tau^{ij} + \tau^{kj} u^i_{,k}) v_j$$

$$E_{ij} = \frac{1}{2}(u_{i,j} + u_{j,i} + u^k_{,i} u_{k,j}) \tag{8-2}$$

where τ^{ij} are the contravariant components of Kirchhoff stress on the deformed convected coordinate net ($\tau^{ij} = J\sigma^{ij}$, with σ^{ij} being the contravariant components of the Cauchy or true stress and J the ratio of current to reference volume), v_j and u_j are the covariant components of the reference surface normal and displacement vectors, respectively, ρ is the mass density, V and S are the volume and surface of the body in the reference configuration, and $()_{,i}$ denotes covariant differentiation in the reference frame. The boundary conditions are:

$$u_1 = 0 \text{ at } x^2 = \pm A/2 \text{ and } x^1 = 0 \tag{8-3}$$

$$\dot{u}_1 = -V(t) \text{ for } x^1 = B \text{ and } -a_p \le x^2 \le a_p \tag{8-4}$$

where $a_p = 2$ mm is the distance along the specimen axis over which the striker and the specimen are in contact and:

$$V(t) = V_1 t / t_r \text{ for } t < t_r$$

$$V(t) = V_1 \quad \text{for } t > t_r \tag{8-5}$$

The impact loading on the Charpy specimen is modeled using $V_1 = 5 m/s$ and $t_r = 20 \mu s$.

The finite element mesh used in the calculations by Tvergaard and Needleman [4] is shown in Figure 8-12 and curves of force versus imposed displacement for selected cases are shown in Figure 8-13.

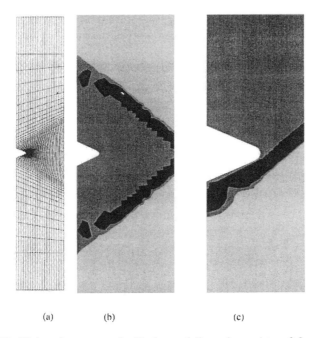

(a) (b) (c)

Figure 8-12: Finite element mesh. Each quadrilateral consists of four "crossed" triangles. (a) Region analyzed numerically, the full mesh has *1480* quadrilaterals. The total number of degrees-of -freedom is *6110*; (b) and (c) near the notch for two configurations; the material in the heat affected zone *(HAZ)* is dark gray and the weld material is the light gray region between the two heat affect zones, adopted from [4].

The force computed in the plane strain calculation is multiplied by the *10mm* Charpy specimen thickness to give the value plotted. Two of the curves are terminated when extensive cleavage occurs and the computations become numerically unstable with the time steps used. Complete brittle fracture can be computed using smaller time steps, but this only changes the work to fracture by a relatively small amount, and is computationally intensive. When ductile failure occurs, the curves are terminated at an imposed displacement, U, of *5 mm*. In all cases, the work to fracture is then computed as the area under curves such as those in Figure 8-13.

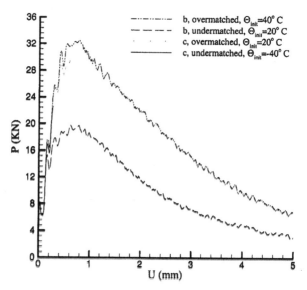

Figure 8-13: Representative curves of force, P, versus imposed displacement, U, for the comparison material properties. $\Theta_{init.}$ is the value of the initial temperature and b and c refer to the weld configurations in Figure 8-12b and 8-12c, adopted from [4].

At $t = 0$, the specimen is assumed to be stress free (so that any effect of residual stresses is ignored) and to have the uniform initial temperature $\Theta_{init.}$.

The effect of weld strength undermatched or overmatched is investigated by Tvergaard and Needleman [4] for a comparison material and analyses are also carried out based on experimentally determined flow strength variations in a weldment in *HY100* steel. The flow strengths of the base and *HAZ* materials are taken to be $\sigma_0^{base} = 930\,MPa$ and $\sigma_0^{HAZ} = 1674\,MPa$. For a 50% undermatched weld $\sigma_0^{weld} = 465\,MPa$, while for a 50% overmatched weld $\sigma_0^{weld} = 1365\,MPa$. The flow strength values used for the *HY100* steel are $\sigma_0^{base} = 790\,MPa$, $\sigma_0^{weld} = 890\,MPa$ and $\sigma_0^{HAZ} = 1140\,MPa$.

For the symmetric configuration 8-12b, the transition temperature for the *50%* undermatched weld is about *-110°C*, while it is about *+20°C* for the *50%* overmatched weld. However, in the ductile regime the work to fracture is higher for the overmatched weld than for the undermatched weld. The transition temperature difference and the difference in ductile work to fracture are both direct outcomes of higher flow strength of the overmatched weld.

For the weld configuration in Figure 8-12c, fracture takes place by cleavage over the entire temperature range. The reason for this brittle behavior is that for this weld geometry, the region of enhanced triaxiality overlaps the *HAZ*. At the higher temperatures, the work to fracture is greater for the overmatched weld because the higher flow strength leads to greater plastic dissipation.

The best understanding of the weld toughness is obtained from curves as those in Figures 8-14 and 8-17 showing the work to fracture as a function of the location of the notch relative to the welded joint. With the notch placed well outside the weld fracture behavior of the base material is modeled, and in the cases studied by Tvergaard and Needleman, where the notch is small relative to the region of weld material, a centrally placed notch gives a good measure of the weld material fracture behavior. Testing Charpy specimens with notch locations near the *HAZ* gives a good impression of the embrittlement resulting from this thin layer of hard material. It is noted in Figure 8-14 for the comparison material that the worst location of the notch is not exactly the same in the overmatched and undermatched cases.

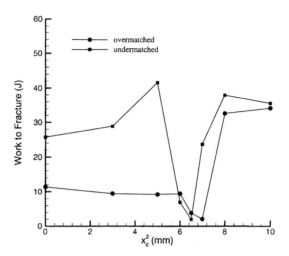

Figure 8-14: Work to fracture versus the location of the *HAZ* relative to the notch as measured by the parameter x_c^2 for the comparison material properties at initial temperature $\Theta_{init} = 0°C$, adopted from [4].

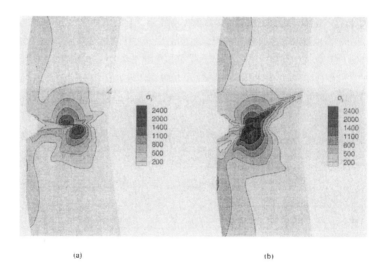

Figure 8-15: Contours of maximum principal stress σ_1. Undermatched weld, $\Theta_{init} = 0°C$, comparison material properties, at $U=2.62$ mm. (a) $x_c^2 = 3mm$, (b) $x_c^2 = 5mm$, adopted from [4].

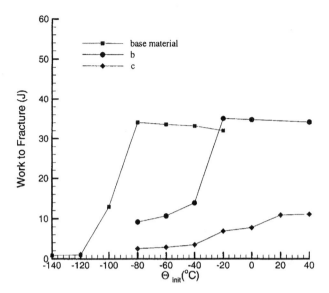

Figure 8-16: Work to fracture versus initial temperature for the *HY100* material properties for the two weld configurations in Figure 8-12 b and c. For comparison purposes, the curve for a Charpy specimen made of the pure base material is also shown, adopted from [4].

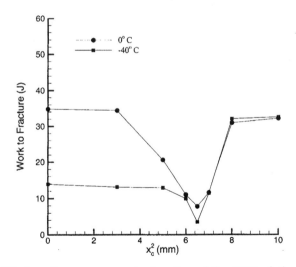

Figure 8-17: Work to fracture versus the location of the *HAZ* relative to the notch as measured by the parameter x_c^2 for the *HY100* material properties at $\Theta_{init} = 0°C$ and $-40°C$, adopted from [4].

Fracture in the Charpy *V*-notch test for a welded joint has been analyzed by Tvergaard and Needleman based on a material model that accounts for ductile failure by the nucleation and growth of voids to coalescence as well as cleavage failure. Pervious ductile-brittle transition studies for homogenous Charpy *V*-notch specimens have shown that this material model is well suited to predict the transition temperature as well as the more brittle behavior under impact than in slow bending, which results from the material strain-rate sensitivity.

8.3 Discussion

Hot cracking is important in certain classes of alloys. This requires models of solidification, including segregation and texture modeling. The stress analysis near the melting point could be the most severe test constitutive models encounter. The yield strength approaches zero, texture and grain boundary effects are becoming significant. Experimentally, it is a hostile environment requiring high speed measurements. The cracking is tightly coupled to the microstructure. It could involve capillary driven flow into thin-walled cracks. Strain in the solid is thought to be a driving force. In this temperature region, viscous or creep effects are expected to be dominate. This challenge could be the most difficult in that it arises from coupling between physical phenomena with length scales ranging from atoms to meters and time scales ranging from nano-seconds to seconds. Mathematical techniques such as operator splitting and Domain Decomposition can be useful in dealing with these disparate length and time scales. To manage this complexity in software is a hard problem in software engineering, [1].

The finite element models developed by Feng et al's study are not intended to and in fact do not reveal the microscopic deformation processes, although they could provide the necessary boundary conditions for the microscopic analysis. Rather, they are used to seek the relations between the macroscopic thermo mechanical conditions local to the trailing edge of a weld pool and the variables that are practically changeable during the welding operation (welding parameters, joint design, application of restraint and so on). The fact that the simulation results correlate well with the experimental observations provides strong support for such an approach.

Dangerously misleading results obtained by conventional Charpy *V*-notch *(CVN)* testing of narrow zones in electron beam welds in an *X-70* fine grained micro-alloyed steel. In one case these results suggested a 50% fracture appearance transition temperature *(FATT)* of *-37 °C (-38 °F)* in the weld metal when a more careful investigation using a specially devised test determined the *FATT* was actually *+90 °C(162 °F)* higher at *+50 °C (122 °F)*. It is proposed that this discrepancy arose because the yield strength of the weld metal was much higher than the yield strength of the base metal. Consequently, general yielding began in the base metal and exceeded the fracture strain of the base metal before brittle fracture initiated in the weld metal. Under these conditions, it is suggested that highly brittle weld metal can yield *CVN* data showing base metal toughness providing the weld metal yield strength is sufficiently high.

The strong variations of the flow stress in a welded joint, with big differences between the base materials, the weld material and the *HAZ*, is usually mapped out by carrying out a large number of micro-indentation tests on a cross-section of the weld. In fact, the yield stresses used for *HY100* in the Tvergaard and Needleman investigation were determined experimentally in this manner. Tvergaard and Needleman studies result, that the fracture toughness of the weld is a more complex property, which not only depends on the flow stress distribution, but also on the fracture properties of each of the material components and on the weld configuration.

It should be noted that the requirements of the material model are higher if the zone near the melt for hot cracking is of particular interest. This improved material model should, at the same time, be matched by a refined spatial and temporal discretization. The modeling of material behavior at higher temperatures and in the presence of phase transformations is perhaps the most crucial ingredient in successful welding simulations. Progress in this field is dependent on the collaborative efforts from computational thermodynamics and material science. The research community active in the field of welding simulation should focus on this field. The improvement in material modeling and increasing availability of material parameters will, in combination with the computational development, increase the industrial use of welding simulations, [8].

References

1. Goldak J.A., Breiguine V. and Dai N. Computational weld mechanics; A progress report on ten grand challenges, International Trends in Welding Research, Gatlinburg Tennessee, June 5-9 1995
2. Chihoski Russel A. Understanding weld cracking in Aluminum sheet, Welding Journal, Vol. 25, pp 24-30, Jan. 1972
3. Goldak J.A., Breiguine V., Dai N. and Zhou J. Thermal stress analysis in welds for hot cracking, ASME Journal of Pressure Vessel Technology, Jan. 24, 1996
4. Tvergaard V. and Needleman A. Analysis of the Charpy V-notch test for welds, Engineering Fracture Mechanics, Vol. 65, pp 627-643, 2000
5. Goldak J.A. and Nguyen D.S. A fundamental difficulty in Charpy V-notch testing narrow zones in welds, Welding Journal, April 1977
6. O'Donoghue, Kanninen M.F., Leung C.P., Demofonti G. And Venzi S. The development and validation of a dynamic fracture propagation model for gas transmission pipelines
7. Feng Z, Zacharia T. and David SA. On the thermo mechanical conditions for weld metal solidification cracking, The Institute of Materials, Mathematical Modeling of Weld Phenomena 3, Ed. By Cerjak H., 1997
8. Lindgren L-E. Finite element modeling and simulation of welding Part I Increased complexity, J of Thermal Stresses 24, pp 141-192, 2001
9. Lindgren L-E. Finite element modeling and simulation of welding Part II Improved material modeling, J of Thermal Stresses 24, pp 195-231, 2001
10. Graville Brian A. Cold cracking control in welds; Published by Dominion Bridge Company, Montreal Quebec Canada, 1975
11. Nguyen D.S. (1975). Effects of heat input on electron beam welds in a C-Mn-Mo-Nb steel. Master's Thesis, Carleton University.
12. Andersson, BAB. Thermal stresses in a submerged-arc welded joint considering phase transformations, J Eng. & Tech. Trans. ASME, Vol. 100, 1978, pp 356-362
13. Jonsson M., Karlsson L. and Lindgren L.E., Deformations and stresses in butt- welding of large plates, in R.W.Lewis (ed.), Numerical Methods in Heat Transfer, Vol. III, p 35, Wiley 1985

14. Chihoski Russel A. Expansion and Stress Around Aluminum Weld Puddles, Welding Research Supplement, pp 263s-276s, Sep. 1979
15. Chihoski Russel A. The character of stress fields around a weld arc moving on Aluminum sheet, Welding Research Supplement, pp 9s-18s, Jan. 1972
16. Matsuda F., Nakagawa H., Nakata K. Kohmoto H. And Honda Y. Trans. of JWRI, Vol. 12, No 1, pp 65-72, 1983
17. Lin W, Lippold JC and Baeslack WA. Welding J Vol. 72, No 4, pp 135s-153s, 1993
18. Arata Y, Matsuda F, Nakata K. and Sasaki I. Trans. Of JWRI, Vol. 5, No. 2, pp 53-67, 1976
19. Matsuda F. and Tomita S. Quantitative evolution of solidification brittleness of welded metal by MISO technique, Proceedings International Trends in Welding Research, Gatlinburg Tennessee, June 1-4 1992
20. Needleman A. and Tvergaard V. A micromechanical analysis of the ductile-brittle transition at a weld, Eng. Fract. Mech., Vol. 62, pp 317-318, 1999
21. Tvergaard V. and Needleman A. Effect of material rate sensitivity on failure modes in the Charpy V-notch test, J. Mech. Phys. Solids, Vol. 34, pp 213-241, 1986
22. Tvergaard V. and Needleman A. An analysis of temperature and rate dependence of Charpy V-notch energies for high nitrogen steel, Int. J Fract., Vol. 37, pp 197-215, 1988

Chapter IX

Input Data for Computational Welding Mechanics

9.1 Introduction and Synopsis

This section describes the author's preferred input data for a Computational Welding Mechanics analysis of a welded structure. Often some desired input data will not be available. In that case one must use the best available approximation that one can find. In addition, one can do analyses with different values of uncertain data in an attempt to bracket the most probable behavior.

9.2 Structure to be Welded

The structure to be welded has a set of parts. Each part has geometry and a material type, e.g., *HSLA x80* steel. (Initial conditions for the thermal, microstructure and stress analysis are discussed with each solver.). The geometry of each part can be defined by a stereolithographic *(STL)* file generated by a *CAD* system or by a parameterized object. For example, a straight pipe can be parameterized by the values of the end points of the axis and its inner or outer radius and wall thickness. Each parameterized object requires a set of parameter names, values and dimensions. For example, ((part Type pipe) (start Point (x_1, y_1, z_1) *meters*), (end Point (x_2, y_2, z_2) *meters*), (outer Diameter *0.6 meters*), (wall thickness *6.2 mm*)). Any object that can be parameterized, can be defined by such a list of parameters specified by names, values and dimensions.

When generating *STL* files from a *CAD* system, options should be chosen so that the *CAD* system preserves the position of the *STL* file for the part in space.

Each weld joint has "ribbon" that defines a curvilinear coordinate system along which the weld procedure is swept. The ribbon could be represented as an ordered set of pairs of points that represents two flow lines that have a start point, end point and hence imply an oriented distance along the ribbon.

Each weld joint has a weld procedure to be described below. A weld procedure can describe a multipass weld.

The order of each weld pass must be specified, i.e., its start time and start position and either or both its end position or direction. We prefer to specify time by year month day hour minute second. For example, *20031104.134522* the year is *2003,* the month *November,* the day *4th*, the hour *13,* the minute *45* and the second *22.* We prefer both welding direction and weld pass end position because it provides a redundancy that can be checked for consistency. A circular weld that starts and ends at the same point must specify a welding direction, e.g., positive or negative direction.

The geometry of fixtures must be defined. Fixtures can either be rigid or compliant. If compliant, the material type of the fixture or its Young's modulus and Poisson's ratio must be specified. This can usually be done simply by specifying the alloy type of the fixture. If the fixture is compliant, the fixture itself usually must be constrained in some way to resist loads. The contact condition between the structure being welded and the fixture must be specified. It could be a contact element that is very stiff in compression and very soft in tension. If the structure is supported by soil, then the soil could be considered a fixture. Soil could have an appropriate constitutive model, such as a CAP model. The soil constitutive model is a function of the soil type, i.e., sand, clay, etc.

The direction of gravity must be defined for this structure, e.g., *(0,-9.8, 0)* means that *+y* is vertically up.

The environment in which welding is being done should be specified. Examples include air, velocity (wind) (v_x, v_y, v_z), temperature *20°C*, pressure 1 bar, humidity *80%*. For underwater welding, air would be replaced by water.

Some parts of the structure could have a special environment, e.g., a pressurized natural gas pipeline could specify the internal environment by methane, velocity (v_x, v_y, v_z) or flow rate *kg/s* or *m³/s*, temperature *20°C*, pressure *60 bar*.

9.3 Weld Procedure

Associated with each weld pass in each weld joint is a weld pool. The weld pool may have any shape and size but its geometry must be specified as input data. If someone has a particularly complex weld pool, they could provide an *STL* file written by a *CAD* program. For many arc welds, an adequate approximation of the geometry of the weld pool can be provided by a half sphere, actually a half ellipsoid, with the length of each semi-axis specified independently. For a half sphere there are five semi-axes. For a Tee-joint, one might choose to describe the geometry of the weld pool by a *3/4* sphere or ellipsoid. Then it has six semi-axes and two of the eight octants of the sphere are missing. The weld pool moves along the weld joint from a specified starting point to a specified end point. The weld pool is always oriented by its position on the weld joint "ribbon" and the welding direction on the ribbon.

For each weld pass, filler metal is usually added. Associated with each weld joint, a mesh will be made automatically that includes joint preparation, weld pool and weld reinforcement. The joint preparation can be specified as an ordered set of *2D* points, in effect line segments, for each 'side of the weld joint'. We prefer the points be ordered *CCW* when looking in the welding ribbon direction, i.e., if one swims along the curve, the base metal is on the left side. The cross-section of each weld pass can be specified by an ordered set of points. Again we prefer the points be ordered *CCW*, i.e., if one swims along the curve, the weld pool metal is on the left side. Usually the weld mesh will overlap the mesh for the structure as a whole. The automatic meshing code will deal with this issue.

For welds, the values of the welding speed, power and arc efficiency are useful. If the weld is pulsed or if weaving is used, that information should be specified. If an estimate of the power density, thermal flux or mass flux distribution from the arc to the weld pool surface is known, that would be useful information.

A weld pass could have more than one weld pool.

9.4 Thermal Analysis Outside the Weld Pool

For the thermal analysis, the initial temperature of each part must be specified. The interface between the moving weld pool and its complement is a Dirichlet *BC* on the complement. The value of Dirichlet *BC* is the specific enthalpy of solid at the solidus temperature. There is also a weld velocity that is 'parallel' or tangential to the weld joint 'ribbon'.

The thermal solution drives the microstructure evolution and the thermal stress solver. The thermal properties required to solve the thermal problem are:

- specific enthalpy as a function of temperature and microstructure
- specific heat preferably as a function of specific enthalpy but often provided as a function of temperature and microstructure
- thermal conductivity as a function of temperature and microstructure
- latent heats of phase transformation; Each phase transformation can have an associated latent heat

9.5 Microstructure Evolution

The composition of the base metal must be specified. We prefer the format $((Fe, *), (C, 0.12), (Mn, 1.3),...)$. Here $(Fe, *)$ means the balance of the fraction is *Fe*. We have to specify whether composition is in weight % or atomic %. Also the initial microstructure of the base metal must be specified. We prefer a format such as:

(ferrite, *wt% 80, gASTM 2*, comp ((*Fe,**), (*C, 0.12*), (*Mn, 1.3*),..))
(pearlite *wt%20, gMicrons 50*), etc.

The composition of the weld metal is a function of the mixing of filler metal with base metal.

For each phase, we suggest the name of the phase and for each attribute of the phase the name of the attribute, the value of the attribute and its dimensions. For example, grain size could be specified by either *ASTM* number (*gASTM*) or diameter in microns (*gMicrons*). Note, strictly speaking pearlite is not a phase as Gibbs would define a phase, but we suspect that most metallurgists prefer to call pearlite a phase. The software should understand that pearlite is actually a eutectic microstructure consisting of lamellae of iron carbide phase and ferrite phase.

For microstructure analysis, we must specify an initial microstructure. If no microstructure is specified, we could assume an equilibrium microstructure with some estimated or default grain size. As a material point moves past the weld or as the weld moves past a material point, the temperature of the material point changes and the microstructure evolves. The latent heats of transformation in the solid state have a small effect on the solution of the thermal problem. The different phases can have different values of thermal conductivity, specific heat and specific enthalpy. For the thermal stress solver, the microstructure determines the specific volume or density as a function of temperature and hence the strain due to thermal expansion and phase changes. It also determines the macroscopic and microscopic elasticity tensor (Young's modulus and Poisson's ratio), yield stress, hardening modulus and viscosity.

9.6 Thermal Stress Analysis Outside the Weld Pool

The initial conditions include initial stress, i.e., the residual stress and initial effective plastic strain for each part in the structure if it is known. If the residual stress in parts before welding starts is not known, it would usually be set to zero. If the manufacturing process for the part is simulated, the residual stress in the part could be computed. The stress analysis can be done as a quasi-static analysis, i.e., inertial forces can be neglected. Both rate independent plasticity and rate dependent plasticity are included. The stress solver chooses its time steps automatically.

The properties required by the stress solver include the elasticity tensor, Young's modulus, Poisson's ratio, yield stress, density or specific volume, viscosity, hardening modulus for rate independent plasticity and softening modulus (a coefficient that defines the rate at which work hardened material softens, i.e., the yield stress decays as a function of temperature). Ideally, these properties should be functions of temperature, time and microstructure. If values of some properties are not available, we suggest using the best available estimates, based on the literature if direct experimental data is not available.

9.7 Weld Pool Solver

If there is a weld pool solver that computes the weld pool free surface, velocity and temperature in the weld pool and weld pool reinforcement, the input data should include the thermal flux, mass flux, current density distribution from the arc and pressure distribution from the arc. The viscosity of the liquid and density of the liquid will be required input data.

If the liquid-solid interface is interpreted as a sharp interface and not a mushy zone, then the jump condition across the interface is the Stefan condition:

$$[q] == q_s - q_l = Lv \cdot n \qquad (9\text{-}1)$$

where q_s is the thermal flux on the solid side of the interface, q_l is the thermal flux on the liquid side of the interface, L is the latent heat of fusion, v is the velocity vector and n is the outward normal oriented from solid to liquid.

The Stefan jump condition implies that the thermal gradient in the front half of the weld pool will be negative.

The melting temperature will be raised on melting and lowered on solidification by the velocity due to kinetics effects.

The melting point will also be depressed by curvature for a solid sphere.

9.8 Material Properties Summary

Properties can be provided for a material or alloy or for a specific phase in a material or alloy. If properties are provided for a specific phase, then the macroscopic or alloy material properties must be computed by a method such as the rule of mixtures, homogenization or a micro-macro model. Properties of a phase are a much more powerful strategy because it allows the properties to be computed as a function of microstructure evolution.

For *CWM* it is best that the properties be functions of temperature or specific enthalpy. Properties can be represented by tables, piecewise polynomials with ranges or can be quite complex functions. Gurson's equation for yield stress as a function of porosity evolution is an example of a complex function.

Internally, the software works only in *SI* units, e.g., *Pa,* not *MPa* and not psi. Of course data could be accepted with other units and then immediately transform to *SI* units. We also transform data or results from *SI* units to other units when post processing and writing reports.

- Latent heats of phase transformation; each phase transformation can have an associated latent heat,
- Specific enthalpy as a function of temperature and microstructure,
- Specific heat preferably as a function of specific enthalpy but often provided as a function of temperature and microstructure,
- Thermal conductivity as a function of temperature and microstructure,
- Elasticity tensor; if a material is not isotropic, e.g., if a texture is present as in some cold rolled steel,
- Young's modulus,

- Poisson's ratio,
- Yield stress,
- Density or specific volume,
- Viscosity,
- Hardening modulus for rate independent plasticity,
- Softening modulus; a coefficient that defines the rate at which work hardened material softens, i.e., the yield stress decays as a function of temperature.

9.9 Visualization

The structure with the following data can be viewed in *3D* color with rotation, zoom and translation:
- Mesh. Part types. Loop over welds,
- Temperature,
- Displacement vector,
- Deformed structure,
- Phase fraction, for each phase,
- Stress. Each component, principal stresses, effective stress, maximum principal tensile stress, hydrostatic stress,
- Strain, Each component, principal strains, effective plastic strain, maximum principal tensile strain, hydrostatic strain, effective strain,
- Internal variables, effective plastic strain, hardening modulus.

9.10 Fracture Mechanics of Welded Structures

For fracture mechanics the location and geometry of the crack or defect must be specified as input data. The crack geometry could be specified as an ordered set of points enclosing the crack surface.

The loading on the structure associated with the crack formation and crack propagation must be specified. If it is the same as the loading on the structure being welded, then the loading need not be repeated. The stress intensity or J-integral could be computed if that was appropriate. We plan to develop fracture analysis based on Gurson's constitutive model or micro-macro constitutive models of fracture that are more appropriate to the heterogeneous material and geometry that is typical of welded structures.

Index